UNLOCKING THE ATOM
A Hundred Years of Nuclear Energy

UNLOCKING THE ATOM
A Hundred Years of Nuclear Energy

Michael Longstaff

FREDERICK MULLER LIMITED
LONDON

First published in Great Britain 1980
by Frederick Muller Limited, London NW2 6LE

British Library Cataloguing in Publication Data

Longstaff, Michael
   Unlocking the atom.
   1. Atomic power — Great Britain — History
   I. Title
   621.48'0941   TK9057

ISBN 0-584-10457-X

Set IBM 11pt Baskerville by Tek-Art, Croydon, Surrey.
Printed in Great Britain by The Anchor Press Ltd., Essex

CONTENTS

ILLUSTRATIONS

PREFACE

My first brush with nuclear energy was in 1924, early in the morning of my ninth birthday when I awoke from a terrifying dream of a bull with gleaming eyes and a loud bellow: I woke to hear the strident ticking of a brand-new alarm clock and to see its brilliantly lit face and hands by my bed. I blamed the dream on the clock and hated it, and from then on I did my best to get rid of it. Many years later it turned up, still glowing almost as brightly as before. A sizeable amount of radium must have gone into the making of its figures and hands. It is probably still glowing at the bottom of a rubbish-tip somewhere.

The next occasion was in 1942 in a German prisoner-of-war camp where some newcomers had brought with them a rumour of work being done on a fantastic new weapon — a bomb 'the size of a Red Cross parcel' (i.e. about ten pounds in weight) that would blast a crater half a mile wide. I, as the only graduate chemist in the camp, was asked what, if anything, this could be. 'Nonsense, most likely,' I told them with all the arrogance of young expertise, but I remembered my alarm clock and I wondered. That night I noticed for the first time that the faint glow from my wrist-watch was not steady but appeared to pulsate as if from a series of little flashes in quick but irregular succession, each one coming from a single atomic disintegration yet giving enough light to be visible to the unaided eye — not a new discovery but one that I had forgotten. Recalling some late pre-war research that I had read about, I wondered still more. Three years later all the world knew the answer.

A few years after my return to England I joined the staff of the United Kingdom Atomic Energy Authority and for the last twenty-four years I have been enthralled with nuclear energy, its history, its technology and the development and implications of its peaceful uses. This book is an effort to impart some of the same interest to other people at a time when, for various reasons (which are discussed in the book) nuclear power is seen by some as a heaven-sent — and possibly final — opportunity for the world and mankind to prosper in harmony as never before, and by others as a threat to man's survival on Earth.

I have not tried to write a scientific textbook, still less a technical manual, or a political tract for or against nuclear power. The book will be found to contain little chemistry, no mathematics or statistics, and (I hope) no politics. I have treated the subject generally on a historical basis and in narrative form, because this is how it came most easily to me. I have covered the development of the first atomic weapons only in so far as this led to the development of nuclear power: on the subject of nuclear weapons, their technology and deployment, I know no more than the next man and have nothing to add to what has already been published on this grim topic.

I have tried to give in simple terms an outline of the nature, manufacture and properties of nuclear fuel and the design of the furnaces or 'reactors' in which it is burned at nuclear power stations — especially those in the United Kingdom because these are naturally the ones which I know most about — and what becomes of it and the waste products afterwards. Safety, which holds a very special place in all aspects of nuclear energy, finds mention in every chapter and also has a chapter to itself, as has a discussion of the economic importance and status of nuclear power, but without setting too much store by detailed — and fallible — forecasts of the future. Other uses of nuclear energy, for example in ship propulsion and in some of the direct uses of nuclear radiations in medicine and industry, are also considered briefly. In the final chapter of the book I try to look forward to the future, both at technical developments including nuclear fusion and at the changing attitudes of people to nuclear power and the significance of this for the future.

Because the subject of nuclear energy is so vast and has so many ramifications, all that I have been able to give is a sketchy account of some of its aspects, and particularly of those that have interested me most. I hope, therefore, that I shall be forgiven for omissions, for brevity and even for superficiality on the one hand, and on the other for over-indulging in some of my own pet interests. For those who want to read further — and I hope they will be many — I have included a short list of books that I myself have found interesting (see p. 171).

All photographs, diagrams and other illustrations (except where stated in the captions) are the copyright of the United Kingdom Atomic Energy Authority, by whose courtesy they are reproduced here.

I have not provided a glossary: all the technical terms used (and I have tried to limit these as much as possible) are explained in the text, mostly at the passages in which they are first used.

I would like to thank my colleagues and friends, too numerous to mention by name, for all the help that they have given me, and my employers the United Kingdom Atomic Energy Authority for permission to express freely views which are not necessarily those of the Authority itself. The responsibility for all that I have written here, whether as fact or as opinion, is mine alone.

Finally I would like to put on record my grateful thanks to my wife, to whom this book is dedicated, for her unfailing help and patience during the twelve months in which the writing of it has been practically my sole spare-time occupation.

Michael Longstaff
Greenwich, 1979

1

PRELUDE TO POWER

Uranium

The story of nuclear power is the story of uranium. It began in the year 1789 with the identification by a German dentist, M.S. Klaproth, of a previously unknown element in pitchblende from the silver mines of Bohemia. Its compounds ranged widely and richly in colour and were found useful in making coloured glass and ceramic glazes: many an antique brown teapot still owes its colour — and perhaps a little radioactivity — to uranium. No other uses were found for it and it remained a mere chemical curiosity having as its major scientific interest its high atomic weight: it was the heaviest known element.

Over a hundred years later, when the age of coal was at its peak and the age of oil had not yet begun, uranium was again holding the attention of scientists. This time the interest lay in its phosphorescence — the glowing colours which could be seen in the dark after the material had been exposed to sunlight or to the X-rays recently discovered by Wilhelm Röntgen. It was in the course of this work that the French chemist, Henri Becquerel, made one of the key discoveries of scientific history. He noticed that some of his plates became 'fogged', apparently by being kept next to uranium compounds without either direct contact or exposure to their phosphorescence.

Closer studies by Becquerel, and later by Marie and Pierre Curie and many others, led to the concept of 'radioactivity' as a property of uranium and, even more strongly, of the other heavy elements that the ore contained. Gradually it became clear that these elements must contain a hidden source of energy many, many times greater than anything hitherto known. During the succeeding forty years the scientific curiosity of enthusiasts working singly or in small groups in university laboratories throughout the world —

and not least in Britain — gradually developed a picture of the structure and nature of matter, down to the level of atoms, that is substantially the same as the picture we have today.

Atoms and nuclei

Looking at the subject in outline we find that all atoms consist basically of a massive positively-charged 'nucleus' with one or more very light negatively-charged 'electrons' orbiting around it, as the planets go round the sun. Every nucleus consists of a tight group of 'protons' which each carry a unit of positive electric charge, and 'neutrons' which have the same weight but carry no charge. (Hydrogen, the lightest nucleus, has a proton only and no neutron.) The protons and electrons balance one another electrically, so an electrically neutral atom must have the same number of orbital electrons as it has protons in its nucleus. If some of the electrons are removed, the atom acquires a net positive electric charge or, if it has extra electrons, a net negative charge; it is then called an 'ion'. When atoms join up into groups or 'molecules' it is these electric charges that hold them together: all chemical reactions, including biochemical ones occurring in living matter, involve electrical attractions between nuclei and electrons. They can be exceedingly complex and are easily upset by the removal of electrons through outside means.

The number of protons in a nucleus is the 'atomic number' and defines the chemical identity of the element, from hydrogen with one proton to uranium with 92 protons. The number of neutrons is more arbitrary and does not affect chemical identity. Atoms having the same number of protons but different numbers of neutrons in their nuclei are called 'isotopes' of the element: they are the same chemically but their atomic weights differ. The number of neutrons runs from zero in hydrogen to 146 in the heavier istope of uranium U-238. The list of elements in this order reveals many striking repetitions of properties and is called the 'Periodic Table'.

In a nucleus the number of protons plus the number of neutrons constitutes the 'mass number' of the atom. Many naturally occurring elements are made up of a mixture of several isotopes having different mass numbers, the average of which is the 'atomic weight' of the element. Unlike mass numbers which are always integers, atomic weights, being averages, are very often fractional numbers.

Radioactivity

In most kinds of atoms the protons and neutrons are bound together very firmly and the nucleus remains stable in the face of any chemical or physical changes going on around it. However, in some atoms — the radioactive ones — the balance of protons and neutrons is wrong, usually through having too many neutrons. As a result the nuclei are unstable and carry surplus energy which they will, sooner or later, get rid of through the process known as 'radioactive decay'. This takes place, not as might be expected through the ejection of one of the extra neutrons, but by changing it into a proton with the forcible ejection of a negatively-charged electron identical with those in orbit around it. In other words a neutron, by ejecting a negative electric charge, becomes a positively-charged proton. The atom will soon pick up the extra orbital electron that it needs to regain its neutrality. For historical reasons the electrons ejected from radioactive matter came to be known as 'beta-particles' and this kind of radioactivity as 'beta-decay'.

Beta-decay changes the number of protons in the nucleus so it leads to a change in the atomic number and chemical identity of the atom: in the case of ordinary beta-decay it goes up one unit in atomic number and one place in the Periodic Table. Sometimes this change is not enough and more energy must be lost before stability is achieved. If the gap is big enough one or more further complete steps may be needed — as many as fifteen in the case of uranium 238 decaying to lead. If beta-decay leaves only fractionally too much energy for stability, the surplus is thrown out immediately in the form of 'gamma-radiation' — a pulse of electro-magnetic radiation energy of the same basic kind as light, X-rays or radiowaves.

Each of the many kinds of radioactive nucleus has its own characteristic degree of stability which is expressed statistically (for we are dealing with very large numbers) in terms of its 'half-life': this is the time taken for half of the atoms in any initial amount of the material to undergo the decay process. Half-lives range from a fraction of a second to millions of years. A short-lived radioactive material will be more intensely radioactive (while it lasts) than an equal number of atoms of a longer-lived material.

Some heavy atomic nuclei, of which uranium is one, contain altogether too many particles to be stable. Sooner or later they take a step towards remedying this by throwing out a small and tightly coherent cluster of particles consisting of two protons and two neutrons. This is called an 'alpha-particle' and the process is known as 'alpha-decay' or alpha-activity. It still leaves an imbalance between protons and neutrons, so it is followed as a rule by beta-decay; in some cases the whole process may repeat itself, accompanied at some points by gamma-ray emission, in a complex 'decay chain' before stability is achieved.

We can now see a possible reason why no element has been found with a heavier nucleus than uranium: anything heavier would have too many particles and would by now have decayed to uranium or something lighter. The question arises, could such heavy elements be made artificially, for example by adding further particles to uranium nuclei? By the late 1930s the experimental techniques, as well as the ideas, were ripe for trying.

Building bigger atoms

One way of attacking this problem was to subject uranium to bombardment with ions of hydrogen or other elements accelerated to great energies by subjecting them to electrical fields in machines called 'particle accelerators' (or more popularly, because of their primary function in atomic physics research, 'atom smashers'). Here it was hoped to use them for building up rather than breaking down, but no evidence of success was produced.

Another approach was to use neutrons which, having no electric charge, would not be repelled by the target nuclei, so that the chances of direct hits and absorption would be greater. It was already known that many kinds of nuclei would absorb extra neutrons, and also that they would do it better if the neutrons were moving comparatively slowly rather than at the very high speeds at which they could be most readily produced. One way of producing neutrons was by bombarding a beryllium target with alpha-particles from a radioactive element such as radium, and one way of slowing them down was by collision with light atoms such as hydrogen or carbon.

Nuclear fission

As early as 1934 Enrico Fermi in Italy had found that neutron bombardment of uranium produced some beta-activity, which was a strong reason for suspecting that a heavier element was formed.

In the mid-1930s several workers had noticed unexpectedly high radiation levels during neutron bombardment experiments, but these were attributed either to alpha-particles or to the notable unreliability of contemporary instruments, and these clues to what later became known as nuclear fission were not followed up. What, one cannot help wondering, would have been the course of history if they had been?

In 1938 two German scientists, Otto Hahn and Fritz Strassmann, after bombarding a uranium target with slow neutrons examined it chemically and spectroscopically for traces of elements chemically akin to radium, which was what might be expected of Element Number 93, the new heavier-than-natural element that they hoped for. But no such chemical was detected. Instead they found distinct indications of elements chemically very like barium – not a heavy element at all but one to be found about half-way up the Periodic Table. They sought confirmation and explanation of this tentative identification from their friends, the Austrian physicists Lise Meitner and her nephew Otto Frisch, both of whom were in Stockholm at the time. Confirmation was quick in coming, together with a bold but plausible explanation: the incoming neutron had caused the uranium nucleus to split into two medium-sized pieces. Closer studies confirmed the reality of this completely unexpected phenomenon, and more detailed experimental work revealed that additional neutrons were liberated at the same time. A word was sought to describe the process and 'fission' was borrowed from the biologists – and kept. The work was published in *Nature* in two letters from Meitner and Frisch in February 1939 and aroused immediate world-wide interest. It seemed to offer a means of releasing at will at least some of the enormous energy already known through the study of radioactivity to be resident in the atomic nucleus, but hitherto only accessible in the world of science fiction.

Further studies revealed more about the actual process of fission (see fig. 1): when a neutron moving with about the same energy as that of the neighbouring atoms – that is, in 'thermal equilibrium' with its surroundings – strikes the nucleus of an atom of U-235 (the lighter and much the rarer of the two major isotopes of that element) it stands a good chance of being swallowed up or 'captured'. This puts the uranium nucleus into a state of such instability or 'excitation' that it can no longer remain in one piece but starts to divide into two by the formation of a narrow and lengthening 'neck', which finally breaks allowing the two sections of the nucleus to fly violently apart, driven by the mutual repulsion of their positive electric charges. In the process two or three neutrons are set free and fly off at great speed, while yet more energy is given off in the form of gamma-radiation (see fig. 1). The two major 'fission product' nuclei pick up their proper quota of electrons and, after repeated collisions with neighbouring atoms during which they raise the general temperature of the surroundings, they settle down as part of the local population of atoms. However, they still contain too many neutrons (even after the loss of neutrons in the fission process itself) and sooner or later undergo beta-decay. Meanwhile the released neutrons collide with neighbouring nuclei and are eventually absorbed. If the fission process takes place in a mass of uranium it is possible that some of the released neutrons may strike other U-235 nuclei and cause further fission, thereby initiating a 'chain reaction' in which a very large amount of energy will be liberated as heat and radiation. This will be more likely to happen if some means is introduced by which the neutrons can first be slowed down, e.g. by collision with nuclei of masses comparable to their own (of which hydrogen of mass 1 could be expected to be the best).

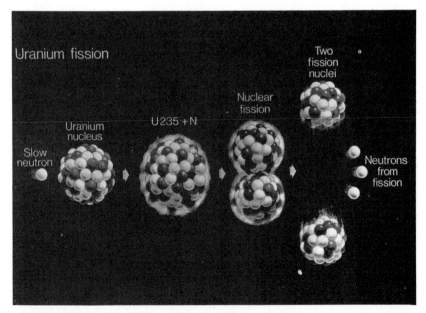

Fig. 1   *Fission of uranium nucleus by a slow neutron.*

Where does the energy come from?

To account for the energy released in fission we must look to Albert Einstein's equation relating energy and mass. These two, he showed, are interchangeable, a very little mass being equivalent to a great deal of energy. In the formula $E = mc^2$, E is energy, m is mass and c is the velocity of light (measured in céntimetres per second — this is about thirty thousand million, an enormous number, which has to be squared in the equation). The energy of fission originates in a loss of mass when a heavy atom splits into two lighter atoms. The difference is measurable in terms of atomic masses; the middle-weight elements are, by comparison with the heaviest (and the lightest) elements, very slightly short-weighted — their atomic weights as determined by chemical or physical means being less than their mass numbers as determined by counting the neutrons and protons in their nuclei. This small 'mass defect' corresponds precisely with the energy liberated in fission (taking into account the free neutrons). It can be looked on as the 'binding energy' needed to hold the nucleus together, and some of it is set free in fission because intermediate atoms need less of it in proportion to their size than do very heavy or very light ones.

If the amount of uranium in which the process is taking place is small, neutrons will escape as fast as they are produced and the chain reaction will peter out, while in a larger mass, where the surface area is smaller in relation to the volume, they will not be able to escape so fast and the chain reaction will grow rapidly in intensity and with it the rate of energy release. If on average just one neutron from each fissioning uranium atom goes on to cause one further fission, 'criticality' is said to be achieved and there will be a steady release of energy. The 'critical mass' of uranium in any system will depend on the precise nature, amount and arrangement of all the materials present in the system. The concept is of fundamental importance in harnessing nuclear energy.

Oddly enough, as a stepping-stone to a better understanding of the nature of matter, the process of fission in itself has not proved particularly fruitful although it has provided some extremely valuable tools of research. However, the practical implications of fission, for peace or war, and indeed for the whole future of the human race were, and still are, immense. Interest was immediate and world-wide. But almost as suddenly as it arose, interest in fission as reflected in the scientific literature and the popular Press died right away. By early 1939 it was clear that war was likely and the work went underground. Here it developed in progressively deeper secrecy, to emerge again in August 1945 in the blast and the toadstool clouds of the bombs that ended the war in the Far East and changed the course of history.

2

WARTIME

The race for the bomb

From 1939 what came to be known as 'atomic research' ('nuclear research' would have been a better term because it had to do with nuclei rather than with whole atoms) stopped being an innocent quest for knowledge, carried out in the quiet of university laboratories by scientists wanting only to find out and tell their fellow-scientists a little more about the nature of things: it became a race in deadly earnest to make a weapon that would bring almost certain victory to the side that got it first, and unprecedented destruction to the runner-up. Secrecy was drawn down like a blind over all the work in this field, both in the countries of the Rome-Berlin axis and those of the Western Allies. Even before war started, a good number of scientists, some of them of great eminence, had fled from the persecutions and restrictions of Nazi-controlled countries, bringing their talents to less hostile shores. The list of names that became famous when the full story at length came to be told was rich with those of German, East European or Italian origin.

The story of wartime research in the USA — the 'Manhattan Project' — and in Canada and Britain leading to the bomb is a fascinating one, but here we will look only at some of the specific problems along the way and at how they were tackled, concentrating on those topics that turned out to be important in the post-war development of electric power generation through nuclear fission. The connection is significant both historically and in terms of present public attitudes to nuclear power.

The research effort had a three-fold role: understanding of principles; development of technology and processes; designing and making the bomb itself. It was soon seen that a weapon based on nuclear fission must be made of a material in which nearly all the atoms are capable of undergoing fission when struck by neutrons: in uranium, no matter how pure chemically, only one atom in every 140 is in this sense 'fissile', and these are the atoms of the isotope U-235. All the rest — 99.7 per cent of the total — are non-fissile U-238. If a bomb was to be made most of this would have to be treated like other impurities and got rid of, leaving virtually pure U-235 as the explosive. It was hard enough to separate the isotopes of any element: they are by definition chemically identical and only differ in the mass of their nuclei and in related properties such as density. If U-235 was to be used, some way had to be found of separating out enough of it cleanly from 'natural' uranium.

Clearly, if kilogram quantities of around 95 per cent purity U-235 were to be had, a very large research and development effort would be needed in this field alone. However, in spite of difficulties, this work was initiated at a very early stage — it was in fact already going on in Britain in 1939 — using a method based on the different diffusion rates of a gaseous compound of uranium known as 'hex' (uranium hexafluoride) through a porous membrane. Diffusion rates differ by not more than about one part in 10,000 so the process is a very slow one: it will be described in detail in Chapter 5.

But it soon appeared that U-235 was not the only fissile material, though it remained the only one provided by nature. Further studies of the heaviest elements on the lines that Hahn and Strassman had pioneered revealed that slow neutrons can in fact be absorbed by the nuclei of U-238 atoms to make a heavier isotope, U-239. This, it was found, was radioactive and decayed in a short time by emitting a beta-particle and changing into the 239 isotope of a previously non-existent element, number 93 in the Periodic Table. This was also found to be radioactive and decayed, again by beta-emission, to the 239 isotope of the next higher element, number 94. These two new man-made elements were named, like uranium, after the outermost of the then-known planets, neptunium for element 93 and plutonium for element 94. (It is not true, though it is often alleged, that the name of plutonium was given because of any particularly hellish properties that the element might be thought to possess.)

Pu-239 was also found to be radioactive: it is an alpha-emitter with a half-life of about 24,000 years. But far more significantly it was also found to be fissile. Furthermore, fission of Pu-239, like that of U-235, resulted in the release of two or three new neutrons and therefore had the potential for supporting an energy-releasing chain reaction.

Plutonium is two places away from uranium in the Periodic Table of the elements, so its chemistry is, on the face of it, likely to be different enough from that of uranium to allow conventional chemical separation techniques to be used for its isolation from the parent material. But to make plutonium requires neutrons and the experimental work that led to its discovery had only required laboratory quantities, such as were available from the action of alpha-particles from radium or polonium acting on beryllium nuclei, or from beams of accelerated ions acting on other suitable target materials. Since the plutonium would be wanted in quantities of around 10–20 kilograms (possibly more) for each bomb, these methods were totally inadequate to meet this objective: apart from other considerations, there was simply not enough radium. It became increasingly apparent that the only likely source of enough neutrons would be from the fission of comparable amounts of U-235 itself.

This looked like coming back to square one, but it was not so. Although a bomb would need nearly pure U-235, obtainable only by isotope separation, there was reason to suppose that given the right conditions a controlled chain reaction — though not a bomb — might be got going at a much slower rate in a mass of 'natural' uranium containing the God-given ratio of 1 : 139 of the two isotopes. This might produce enough neutrons to convert a small proportion, perhaps a few parts in a thousand, of U-238 into plutonium. Because of the analogy with the reaction-vessels or 'reactors' used in the chemical industry, such an assembly of uranium and other hardware needed to do this job came to be known as a 'nuclear reactor'. The neutron bombardment process was called 'irradiation'.

The two routes

So there were two possible routes to the bomb — U-235 from natural uranium via isotope separation, or plutonium to be made from U-238 by irradiating it in a reactor in which controlled fission of U-235 provided the neutrons. Some of the research work was common to both routes, some was concerned with one route only. A substantial fraction dealt with weapons as such, but this we are not concerned with here. Let us look at some of the problems that came up and were, for practical purposes, solved during the war-time research, but only in so far as these related to the later uses of nuclear fission for peaceful purposes. The most important work by this criterion concerned the design and operation of nuclear reactors which subsequently would become the furnaces of the nuclear power industry, and uranium which was to be their primary fuel. As we shall see, plutonium, produced from uranium, became a secondary fuel allowing indirect use to be made of the plentiful but otherwise useless U-238. The isotope separation route, though outwardly very much the simpler, proved difficult and expensive: it has been important, though not essential, in the development of nuclear power.

The fission process

The whole project depended on fission. This had only been achieved by a handful of workers and firm verification, both of the experimental work and of the proposed explanation, was lacking. This being achieved, details were needed: What speed of neutrons gave the best results? What were the chances of a neutron hitting a U-235 nucleus in passing through a mass of uranium? If a hit was scored, what were the chances of fission taking place, or of some other interaction? For example, it might bounce harmlessly off, or be absorbed and form an atom of U-236. What was the average number of neutrons produced per fission — 'sometimes two and sometimes three' was not good enough — and with what energies did they arise? Precisely what fission products were produced, and on average how much of each? With what energies did they move? Were they radioactive, and if so, what were their half-lives? Did any of them tend to swallow up neutrons so as to interrupt the chain-reaction of fission? How did the energy of a neutron affect its behaviour in this respect? All these questions had to be, and were, answered.

Uranium, the key element, was at that time significant chiefly as an unwanted constituent of radium ore, but now it became suddenly important. All available stocks of uranium-rich ore, and of slags and wastes from radium extraction, were bought up by or for the governments of the Western Allies. Intensive research was undertaken, both into the practical chemistry and the related technology of the element and its ores, and into the deeper reaches of its chemistry and physics. Chapter 5 describes uranium and its role in nuclear power in some detail.

What had to be done for uranium also had to be done for other key substances. This involved the obvious ones such as plutonium and the materials likely to be used in nuclear fuel and in the structure and operation of nuclear reactors, isotope separation plants and other chemical plant, also the whole range of fission product elements, including some that were previously unfamiliar or even non-existent.

Coping with radiation

Radioactivity and X-rays had been known for over forty years before fission was discovered and a great deal was already understood about their damaging effects on living matter: much had been found out the hard way by scientists working with radioactivity, several of whom had died from radiation-induced cancers. Radiation effects had become the subject of much study, but a great deal had still to be discovered and applied to the work in hand. Accordingly, the new interdisciplinary science of 'Health Physics' grew, from its beginnings in a few hospitals, to an all-pervading importance at every stage and level of nuclear energy research and applications. Chapter 9 covers in detail some of the problems that radiation gives rise to.

Uranium itself is only very slightly radioactive, but in its decay it gives rise to a whole succession of radioactive 'daughter-products', some of which (notably radium) are even more unstable than the 'parent' and decay much more rapidly to the next stage. Uranium ore, and the materials left after the uranium has been extracted, are more strongly radioactive than the uranium itself and require more care in handling.

When uranium undergoes fission not only is energy given off as heat, but there is also a burst of gamma-radiation from the fissioning nucleus, while the fission products themselves are radioactive, each kind beginning to decay at its own natural rate. Furthermore, the neutrons given off in fission, besides constituting a dangerous form of radiation in their own right, can cause adjacent materials to become radioactive. Plutonium is also radioactive — more so than uranium — and it too gives rise to fission products and neutrons and to its own range of radioactive daughter-products.

Ways had to be developed to do the work in spite of the radiations, none of which can be detected by any of the human senses, so they give no direct warning of their presence. Instruments therefore had to be devised and put into operation to monitor and record levels of radiation and to give unmistakable warning if these got too high. Workers had to know and understand the rules, and procedures had to be laid down, and enforced, for their observation.

Let us look at the main features of the problem. First, radiation can shine from its source onto a person, and to a greater or lesser extent into his body. If the man leaves the area, or if the source of radiation is taken away or suitable shielding is interposed, the hazard disappears. Secondly, radioactive substances may get onto the man's skin or clothes where they will be carried around with him until he changes his clothes or, by washing, removes the source of radiation; it may be on surfaces such as floors, walls or benches where he works. This is called 'contamination', which is a long word meaning 'dirt', and the answer to it is to clean up or 'decontaminate' — which may be tedious, difficult and even unpleasant. So steps are taken to try to keep contamination levels, whether of the person or of the working area, to a minimum by confining radioactive substances strictly to where they are meant to be. Thirdly, radioactive materials may be taken into and lodge within the body where they can do intimate and direct damage to the living tissues, and from which they can often be very difficult to remove.

Where penetrating radiations — gamma-rays and neutrons in particular — are present, 'biological shielding' is installed to prevent the radiations from reaching the workers, in the same way that extraneous light must be shut out of a photographic dark-room lest it fog the films. The nature and thickness of the shielding will depend on the radiations, and may vary from a few millimetres of transparent plastic for 'soft' beta-radiation to 2—3 metres of concrete or 5—6 metres of water for high energy gamma-rays, X-rays or neutrons.

The need to work through, around or over shielding of this nature led to the development of remote-handling equipment ranging from simple tongs to more elaborate 'through-the-wall' tools, often operating by means of a heavy ball-and-socket joint: even more complex 'master-slave manipulators' were invented which faithfully reproduce, at a distance of many feet and on the far side of a considerable thickness of concrete, the most delicate movements of the operator's hands (see fig. 2). Viewing may be through thick glass blocks or through several feet of a dense liquid such as zinc bromide, or by closed-circuit T.V.

It is only when the danger lies in contamination of the clothes or body surface by loose radioactive materials, or the inhalation or ingestion into the body itself, that protective clothing (other than

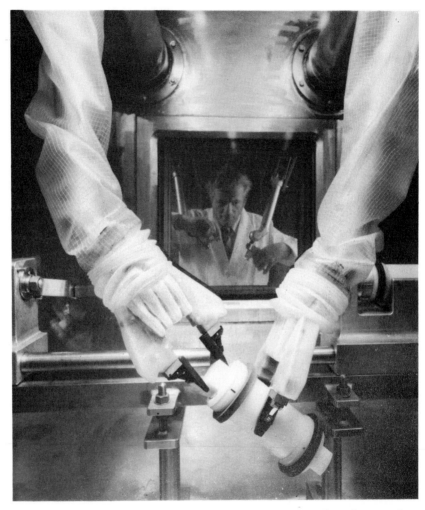

Fig. 2   *Examination of a highly radioactive process sample, using a modern master-slave manipulator. The operator is protected by some 3 feet of concrete, with a viewing window of liquid zinc bromide.*

simple overalls) needs to be used. The 'spaceman' or 'frog-suit' type of clothing beloved of science fiction writers was developed as a method of ensuring that clean air is breathed and the body is not contaminated. It gives protection against personal contamination by radioactive materials, and against alpha-particles, but it gives no protection at all against beta- or gamma-radiation (see fig. 3). For

Fig. 3   *Protective clothing and piped air supply used in an area where radio-active dust may be present.*

most kinds of work not needing heavy shielding it is much simpler to box the radioactivity up and leave the operator free, by using 'glove-boxes' or small self-contained experimental enclosures set up as 'mini-labs' or 'mini-workshops' into which the operator's gloved hands can penetrate through sealed ports. (see fig. 4).

The need for all these precautions against radiation enhances the difficulty, reduces the pace and increases the cost of nuclear research. It is also one of the causes of the awe and mystery that still surrounds it in the public mind. Nevertheless, it is a necessary and amply worthwhile expense.

Plutonium

Let us remember that there were two possible routes to the bomb — uranium 235 or plutonium. The U-235 route is of little further concern to us here: we will return to it to examine the process of isotope separation or 'enrichment' when we consider fuel for nuclear reactors. The plutonium route, on the other hand, had of necessity to include the design and construction of the very first nuclear

Fig. 4   *An up-to-date laboratory equipped for the analysis of plutonium carbide for future fast reactor fuel. Argon gas at reduced pressure replaces air in the glove boxes.*

reactors in which a self-sustaining chain reaction of nuclear fission took place. As such it is a vital part of the story of nuclear power. Let us look at what the task called for and how it was achieved.

First, uranium was needed as pure as possible and in sufficient quantity to build up a 'critical mass' — that is to say enough to keep the supply of fresh neutrons from the interior ahead of their loss from the surface. (This involved calculations of great difficulty that had to be made on a basis of quite inadequate knowledge, but that nevertheless got very close to the right answer.) Second, a means was needed of slowing down the newly-formed fast neutrons to speeds at which fission would be readily caused in the comparatively rare U-235 nuclei: graphite was chosen for this important task, because carbon atoms allow neutrons to rebound freely from them and are light enough to slow down ('moderate') the neutrons effectively. Third, a method of controlling or stopping the chain reaction was needed, by removing neutrons from circulation. Boron was one of the elements found to have a powerful capacity for capturing neutrons, and bars of boron steel were used as 'control rods' which could be inserted and withdrawn at will.

The world's very first nuclear reactor was built in, of all places, a squash court on the campus of Chicago University under the personal direction of the refugee Italian physicist, Enrico Fermi. It achieved criticality — that is to say it worked for the first time — in December 1942, laying the foundation for most of the world's nuclear weapons and all its nuclear power stations. It proved that a self-sustaining and controllable chain reaction could be made to take place in 'natural' uranium, that such a reactor would produce enough neutrons to turn a small proportion of the U-238 into plutonium, and that it would get hot. Bigger and better reactors followed in direct line of descent from Fermi's 'Chicago pile', culminating in the huge plutonium-producing reactors at Hanford.

The next step was to isolate the plutonium, and here firm information was not just in short supply: it was, like plutonium, non-existent. However, knowing the position of plutonium in the Periodic Table of the elements and hence the number of electrons in orbit around it, and working on the assumption that they would be arranged in a way that continued the patterns already seen in the Table, it was possible to make reasonable guesses as to what some of the chemical and physical properties would be.

Gradually it became possible to build up very detailed information about its chemical, physical and mechanical properties, and later those of the other 'transuranic' elements. Separation processes were devised for operating at the scale needed to produce highly purified plutonium in kilogram quantities. These processes, much modified and improved, still form the basis of the nuclear fuel reprocessing operations associated with nuclear power generation today — although for reasons of official secrecy in the post-war days each country has had to re-invent most of the technology for itself.

The two routes to the bomb — uranium 235 and plutonium — converged in the summer of 1945 as the materials, the designs and the hardware for both weapons all became ready. Because of the different nuclear characteristics of the two explosives, the designs of the bombs using them differed, even in size and shape. There seemed to be almost a neck-and-neck race for the first to be ready and both were in fact finished within a few days of each other. But the fission-bomb principle needed testing in practice before it could be said to be ready for use, and in July 1945 an explosive device was set up at Alamagordo in the New Mexico desert, having all the essentials for a test explosion of the smallest size that would work. A few kilograms of plutonium was assembled in the form of two hollow hemispheres that could be blasted together into a single more-than-critical mass so that a very rapid chain reaction would proceed through the whole mass before it disintegrated in an explosion of unprecedented power.

The test took place on 16 July 1945, and succeeded beyond the hopes — or fears — of all involved. So did the two real bombs — one of plutonium and one of U-235 — dropped on the two Japanese cities a fortnight later. The war ended within days.

3

THE RUN–UP
TO NUCLEAR POWER

After the war

The bombs dropped on Japan not only ended the war: they introduced 'the Atom' – nuclear energy – to the world in the most dramatic way possible. They also brought to an end the wartime collaboration between the Allied teams of USA, Britain, France and Canada. Just when it looked as if the occasion had arisen for a real collaborative effort to turn this new force away from warlike purposes and to the benefit of man, the hard facts of international power politics intervened. The United States Congress made it clear that there must be no chance of the secrets of US atomic research, past, present or future, going astray. The McMahon Act became law in August 1946 and, in spite of earlier promises, collaborative research ceased forthwith. From then on, in atomic work, it was each country for itself.

Britain and France set out independently to make their own weapons. Canada continued to act as host to a British research team, but renounced making the bomb and concentrated on peaceful applications. The United States got on with the job of making better bombs and the means of their delivery, and began work on the far more powerful 'H' bomb, based not on fission of heavy nuclei, but on 'fusion' of light ones (deuterium and tritium, isotopes of hydrogen) – the process known to keep the sun and the stars burning. She also looked hard at the possibility of nuclear power as a way of propulsion for warships, and especially for submarines. We will return to these developments in Chaper 11.

Britain and the bomb

In October 1945 the British Government announced an open-ended decision to set up an atomic energy research establishment and, in January 1946, a manufacturing organisation to produce fissile materials, under the control of the Ministry of Supply (Atomic Energy Department). Thus the names of Harwell and Risley began to become known outside their own neighbourhoods, and later became famous as centres of nuclear research and development. Not until May 1948 did the Government announce that Britain was to go ahead with the design and production of nuclear weapons of her own as a matter of urgency.

Although Harwell started under the impetus of military needs and research into the processes and materials applicable to them, the establishment was concerned with fundamental as well as applied research and was never directly involved in the design and production of nuclear weapons. This was the concern of the atomic weapons research team, initially at Fort Halstead and later at Aldermaston. Harwell from the first looked forward to peaceful developments, including the eventual production of useful power. This had to wait, but meanwhile Harwell developed a small peaceful side-line in the use of nuclear reactors as a new and potentially prolific source of radioactive materials, and fostered their use in medical and research applications, and later in industry (see Chapter 11).

If Britain was to catch up with work on the bomb, she now had to pick up the pieces and put together again the jigsaw-puzzle of wartime research. Some of the pieces in the fields in which Britain had been working were clearly identifiable and even getting on to completion in some detail. Others were less clear in their import, or the known results were only fragmentary, while in other areas where the British teams had not been involved at all and had therefore not been informed, there were great blanks in the picture. Not all the gaps could be filled in because it was clear that Britain on her own could not possibly complete the entire picture in any realistic timescale. Decisions then had to be taken on what to concentrate upon and what to leave out. But Britain now had one clear advantage that all the combined forces of the wartime research project lacked: her scientists, like everybody else, knew for certain that a workable bomb could indeed be made. There was no psychological barrier of uncertainty to be crossed. And those who had been working on the project already knew the 'feel' of the work and were familiar with some of the techniques characteristic of atomic energy research: health physics precautions, including shielding and remote handling techniques; the vital importance of cleanliness and good housekeeping throughout; the often unexpected significance of very minor impurities in materials, and the need for new standards of strictness in analytical control; the use of high-voltage machinery; the effects of radiation on chemical behaviour; the physics and chemistry of radioactive materials; and the strange nature of matter at the sub-atomic scale – all these formed part of the background in which the scientists had already been working.

The first major decision that had to be taken, which was to have a profound and lasting effect on the subsequent development of nuclear power in Britain, was whether to make the first bombs from U-235 or from plutonium. It would be clearly foolish to follow two separate paths simultaneously when it was now known that either would reach its objective. The U-235 route was, so it seemed, much the more direct and straightforward of the two: crude uranium oxide would have to be purified, converted to hexafluoride and passed through a multi-stage gaseous-diffusion plant, from which the 'enriched' fraction, reconverted to the metal, would form the required, nearly pure U-235 explosive.

The plutonium route, in contrast, would involve purification of the uranium oxide, its conversion to metal and fabrication into nuclear reactor fuel. It would be necessary to construct a nuclear reactor in which fission would occur as a chain reaction, at the same time creating plutonium in the fuel; this would then be separated out chemically and converted to the metal to form the alternative nuclear explosive.

Considering the relative complexities of the two processes, it is at first sight surprising that the plutonium route was chosen. But the reasons for the choice were overwhelming: from the wartime research it was clear that isotope enrichment, whether by multi-stage gaseous diffusion, or by the possible alternative of electromagnetic separation, required the expenditure of enormous amounts of expensive electric power that Britain (unlike the United States) could not afford. Added to this was the difficulty of preparing the diffusion membrane, of which literally acres would be required: scientists in the British team already knew enough about this at first-hand to be aware of the difficulties. Finally, it was doubtful whether Britain's post-war manufacturing industry was capable of producing the thousands of compressors and ancillary equipment to the very exacting specifications needed for the work.

The plutonium route to the bomb presented fewer difficulties and appeared in every way the more practical course to follow. It would involve the preparation from crude uranium oxide of very pure uranium metal rods, enclosed for protection in a metal of which aluminium or zirconium were already known to be suitable; this would involve chemical and metal-working technology already known to be feasible. A nuclear reactor would have to be designed and constructed in which the fission chain reaction could be carried out with the simultaneous creation of plutonium. This, it was already known, could be done in a reactor using graphite as a moderator to slow down the fission-produced neutrons, taking away the excess heat by water cooling. It would not be unduly difficult to make or procure and machine graphite of suitable quality. On completion of the irradiation in the nuclear reactor, the fuel would have to be dissolved in acid (no difficulty here) and the solution put through a separation process for extracting the plutonium. This also was known to be practical, using a solvent process in which an organic liquid – dibutyl carbitol – was employed. The extracted plutonium nitrate would have to be reconverted to plutonium metal for use in the bomb. A substantial research and development effort would be needed at each stage, but no one step would be impossible, nor did it appear to be either too difficult for the scientists to negotiate or too demanding for British industry to meet. Extensive research, however, to confirm, repeat, re-invent, enlarge and improve upon the wartime work, to break new ground and to look ahead, would all be needed, together with industrial development of quite a new kind, over a period of months or possibly years.

Harwell

With this approach firmly established, the Harwell teams set about equipping themselves for the job. The new tools and paraphernalia of nuclear research – reactors, particle accelerators, remote handling equipment, special electronic instruments, radiation shielding, protective clothing, special chemical plant and much else besides – were unavailable and had to be designed and built specially, often on a hit-or-miss basis, with far too little information to go on. Buildings already on the Harwell site – an RAF airfield – were adapted and services installed, including a greatly enlarged electric power and water supply, and a drainage system which would have to cope with all the foreseeable needs of the site. This would have to include the storage, treatment and eventual disposal of large amounts of possibly radioactive effluent water, to comply with the appropriate regulations when these were finally brought in (see Chapter 8, p. 120).

Clearly, much research equipment would have to be built on the spot, so very comprehensive engineering workshops would be needed. These workshops were housed in one of the existing air-craft hangars and were among the first parts of the establishment to be brought into operation. One of the first new buildings to be erected was to house the radiochemical laboratories where radio-active materials in great variety and at high levels of activity would have to be worked with. The building and the laboratories that it housed were, in effect, built around a specially designed air ex-traction and ventilation system, incorporating many features specially designed for clean working and the avoidance or isolation of radioactive contamination. After thirty years the only major changes that have been needed have been substantial enlargements to the original building. Looking further ahead to the scaling-up to pilot plant or small production installations, a building was specially designed for chemical engineering developments, to accommodate plant that had not yet even been designed or its processes and functions determined. This same building is still in operation, having seen many changes of equipment housed in it.

As the needs of the researchers became clear, equipment was designed and procured, or more often built from such materials as were available, and installed in the buildings as they became ready. Cyclotrons and other particle accelerators (the 'atom smashers' referred to in Chapter 1 which had long been the trad-itional tools of nuclear physicists and were so named from the days when atoms were thought to live up to their reputation of being unbreakable) were among the largest individual items. These machines were designed by and built under the supervision of Harwell scientific staff; their function was partly to provide under-lying scientific knowledge and understanding about the behaviour of atomic particles and their interactions together; partly to determine, as precisely as possible, the ways in which different materials reacted to the impact of particles of all kinds and energies; and partly to study the fission process itself. One of these, a 'synchro-cyclotron' of 7 million volts accelerating power, was closed down in 1979 after nearly thirty years of almost continuous use.

Nuclear reactors

Most vital of all, the earliest possible start was made on the first nuclear reactor, GLEEP (Graphite Low-Energy Experimental Pile).

This was (and is) a simple stack of graphite blocks with horizontal channels to hold the uranium fuel rods and vertical holes to take the neutron-absorbing control rods (see fig. 5). With 32½ tonnes of uranium metal (including initially a proportion of uranium oxide) as its fuel, most of which was supplied by ICI Ltd, GLEEP 'went critical' — that is to say a self-sustaining chain of reaction started — on 15 August 1947, thus becoming the first nuclear reactor to go into operation in Western Europe. (A very similar reactor in Moscow beat it by about six months.) GLEEP's main jobs were first, to help in the study of reactions between neutrons and nuclei by providing, as it were, a bath of neutrons in which specimens of materials (including biological specimens) could be steeped to see what nuclear reactions or other effects were caused. Second, it was to establish the nature and behaviour of one of the simplest of all possible kinds of nuclear reactor. Third (in its earliest years), it provided a limited source of supply of radioactive materials, mainly for medical purposes, a task later taken over by BEPO (see below).

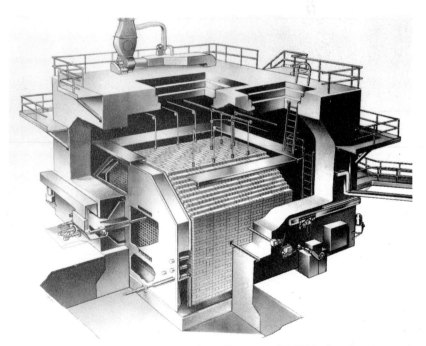

Fig. 5   *Cutaway diagram of the Harwell reactor GLEEP, showing the stack of graphite blocks and the ends of the channels containing the fuel elements.*

Fig. 6   *The research reactor BEPO at Harwell. The fuel rods lie in horizontal channels at right angles to the control rods (left), whilst shut-down rods operate from the top. BEPO operated from 1948 to 1968.*

The nuclear fission reaction in GLEEP goes on at such a low level that it does not take much to stop it altogether: quite low impurity levels in the graphite, for example, would cause it to shut down. For this reason it has proved to be a most useful instrument for measuring neutron-absorbing impurities in nuclear materials, particularly in reactor-grade graphite; and it continues to this day to be the standard 'test-bed' for the nuclear graphite industry. The removal of heat from GLEEP presents no problems — in normal operation only a few kilowatts of heat are produced (though it has been run up to an output of 700 kW).

GLEEP was soon followed by the much larger nuclear reactor BEPO (British Experimental Pile, Zero Energy) which became operational in July 1948 (see fig. 6). The basic principles were the same as in GLEEP, but in BEPO the nuclear chain reaction proceeded with a much greater intensity so that there were many more neutrons available at any point: that is to say the 'neutron flux' was much higher as measured in numbers of neutrons passing through a 1 cm^2 hypothetical target every second. The figure for GLEEP was about one thousand million at the centre of the active 'core', whereas that for BEPO was about one and a half million million. Because the core was so much larger than that of GLEEP (a 30 ft cube as against an 18 ft cylinder, approximately) and because of the more intense nuclear reaction, 6 megawatts of heat had to be removed all the time from BEPO; this was done by powerful blowers which drew filtered air through the fuel channels in the core and discharged it, filtered again, through a 200 ft chimney-stack. The air pressure in the core was kept all the time slightly below atmospheric pressure so that if a leak occurred it would be of fresh air into the reactor core, not of possibly contaminated air from the core to the surrounding atmosphere: this was an early example of a simple but effective nuclear safety measure introduced at the design stage. Much more about safety will be found in later chapters, but it may be mentioned here that BEPO was controlled by neutron-absorbing rods moving in channels at right angles to the channels carrying the fuel elements. These could be operated either manually or automatically and had a further automatic override to shut the reactor down in case of an error by the operator. Reactor control was supplemented by emergency shut-down rods poised over vertical channels and held in place only by electro-magnets: if a power failure, for example, cut the electricity supply to the blowers, the magnets would de-energise and the rods would immediately drop into the core, boosted on their way by a blast of compressed air, and the chain reaction would be halted within a fraction of a second. This provides a further example of the 'fail-to-safety' philosophy that pervades all nuclear energy work. BEPO was closed down in 1968 and its remaining work taken over by the much more powerful materials-testing and research reactors DIDO and PLUTO (see Chapter 4).

There were other areas of research at Harwell which could be planned and at least partly provided for in advance: these included mineral treatment, chemical engineering, the design and building of large-scale plant, metallurgy, electronics and instrument design. These last were linked to the discipline most characteristic of atomic energy research, health physics, and this in turn had close links with medicine, both in the routine care of staff and in research and development work connected with the effects of radiation on living matter.

Risley and the North

At the same time as Harwell was starting up as a research establishment, work was going on in the Engineering Design Offices at Risley in Lancashire. Plans were being made for building the big reactors that would produce the plutonium needed for the weapons programme. Working at first largely from theory and the wholly inadequate data brought back from America, supplemented later by research work at Harwell with GLEEP, BEPO and the particle accelerators, the Risley engineering teams designed a pair of graphite-moderated reactors ('piles'), each of them roughly 50 ft graphite cubes to be fuelled with uranium metal rods sheathed in aluminium cans.

Far more heat would be generated than in BEPO and it was expected that, as in the USA, this would have to be removed by passing large amounts of cooling-water through the piles. But calculations and the records of United Kingdom river authorities revealed a woeful shortage of suitable river sites: flow would have to be large, reliable right through the year, and the water itself would have to be of very high purity. Only one river appeared to meet these criteria — the short stretch of Scottish water between Loch Morar and the sea, a place of great natural beauty. However, heat-transfer calculations and experiments, combined with the measurement of thermal neutron absorption in aluminium, indicated that water-cooling might not be necessary after all: if the aluminium cans were fitted with extended fins along their length, it might just be possible to use air, not water, for cooling. Experiments confirmed this, and two large air-cooled piles were built at a remote Cumberland site, near to the village of Seascale. With their curiously-shaped chimney-stacks (filter chambers at the top were added as a safety precaution after construction had started) they dominated the countryside. The site is now well-known to the world as 'Windscale' (see fig. 7).

Many hundreds of tons of uranium metal had to be provided before the piles could be brought into operation. But hitherto uranium had largely been neglected by the chemists because it had very little industrial use, its sole interest being as an impurity that always accompanied the vastly more valuable radium. The Risley engineers again based their plans partly on the rather meagre and not always accurate data available in the pre-war scientific literature, and partly on such information and experience as had come out of the wartime work in North America. This was now supplemented

Fig. 7   *The Windscale and Calder Hall site as seen from the surrounding country, showing (left) the cooling-towers of Calder Hall, the two stacks of the early plutonium-producing piles with the original chemical plant behind them. On the extreme right is the sphere containing the Windscale Advanced Gas-Cooled Reactor.*

by a rapidly increasing output of research results from Harwell, and on this basis they were able to design and set up a production factory for uranium fuel at Springfields in Lancashire. Here imported uranium ore was converted to fuel to supply the Windscale piles and to keep BEPO in operation at Harwell.

However, the system was not complete: virtually nothing was known about the plutonium which would have to be extracted from the irradiated rods as they came from the piles, because all the work on that vital material had been closely guarded and exclusively kept in American hands during the wartime research. But it was not necessary to go back to the starting-point of the Periodic Table: a tiny quantity of plutonium — under 1 gram — was available at Harwell from Canada and from this enough was learned about the properties of this strange new element to enable the chemists to devise a process for its extraction and purification, and to assess this as a continuous process on a laboratory scale.

Information about uranium and plutonium, however detailed, was not in itself enough: once the uranium rods have been in the reactor, the fission process produces small amounts of a great many other materials — the fission products. These comprise about forty different chemical elements, each with its own properties and its own way of behaviour — and most of them also radioactive. Some of these elements were already familiar — strontium, iodine, caesium, etc. and others less so, for example ruthenium. Yet others hitherto had been laboratory curiosities only, or even, in the case of technetium and astatine, non-existent in nature. Enough had to be found out about their chemistry to devise ways of cleanly separating them from the plutonium and the bulk of the residual uranium.

Radioactivity, which as we have seen already is an ever-present accompaniment of nuclear fission, imposed its own characteristic handicaps. Work involving plutonium necessitated containment in sealed areas to prevent risk of inhalation, while that involving fission products needed heavy shielding against penetrating gamma-radiation. Appropriate shielding, containment and remote handling techniques had to be devised for all stages of the separation process, and this employed substantial research and engineering effort at Harwell and Risley.

Plutonium, being fissile, imposed the added problem of criticality: on no account must sufficient plutonium be brought together to allow a chain reaction to develop of its own accord. So the plant had to be designed to make this virtually impossible, whatever unlikely combination of plant malfunctions and human fallibilities might arise. This again needed a substantial research and design effort.

Early in 1952, the separation plant at Windscale produced the first trickle of plutonium solution, to be reduced to metal for the weapons design teams based on Aldermaston to work on. In October 1952, Britain successfully exploded her first nuclear test-weapon and became, after the United States and Russia, and before France, the third member of the 'club' of the world's nuclear powers. In Britain, and elsewhere, weapons research continued with improvements on the 'A' bomb based on fission, and the development of the far more powerful 'H' bomb, based on fusion of light nuclei.

Success with the weapons programme allowed more thought to be given to the ways in which the new-found source of energy could be used for constructive and peaceful purposes. All through the war and post-war years scientists and others had allowed themselves to day-dream about nuclear power stations, nuclear-driven ships, trains and planes, chemical industries based on nuclear heat or radiation, medical and agricultural advances based on radioactivity, and all the other benefits to man known collectively as 'peaceful uses of atomic energy'. Later these were for a time grouped more formally under the well-meaning slogan of 'Atoms for Peace', a campaign officially inaugurated by President Eisenhower in 1953.

However, by this time Britain had pressed ahead on another front, the separation of uranium 235. A gaseous-diffusion plant had been designed and constructed at Capenhurst in Cheshire, using batteries of rotary compressors and porous membranes designed and developed in this country, and taking electric power for its operation equivalent to the requirements of a city the size of Liverpool — but still small compared with the vast American separation plants. The plant was used for producing virtually pure U-235 for weapons purposes, and for research. Later it was to be used for producing 'low-enrichment' material for developments in the nuclear power programme (see Chapter 5).

Calder Hall

An opportunity now arose that combined the needs of defence with the dreams of peaceful atomic power. Even supplemented by the Capenhurst output of U-235, the two reactors at Windscale, big though they were by the standards of the day, could not generate enough plutonium for Britain's weapons programme. A further source of supply was needed and this meant more reactors. Each of the two Windscale piles was already having to get rid of about 150 MW of heat — enough to keep 50,000 households warm — and it seemed that there should be some way of using this heat, preferably for generating electricity. From the pool of ideas stirring at Harwell as sketches, fantasies and doodles on the backs of envelopes, and as serious studies, one in particular looked as if it really might work; better still, according to available evidence and reasonable assumptions it should produce electricity at not much more than the current market costs of coal-fired generation. This crystallised into the PIPPA project — Pile for Industrial Production of Plutonium and Power. Its authors' claim, with support from a full feasibility study, received Government backing in 1952. Ultimately the project took shape at Calder Hall, alongside Windscale, and became the world's first commercial nuclear power station. It was opened by Her Majesty Queen Elizabeth II in October 1956 (see fig. 8). It was not the first nuclear reactor ever to supply heat to drive a dynamo: that prize had already fallen to the Russians, but only a few kilowatts had been generated and then only for a

Fig. 8   *The opening of Calder Hall by Her Majesty the Queen on 17 October 1956.*

short time. The United States also had built and operated a small reactor which supplied heat to a boiler and turbo-alternator producing some 20,000 kW of electricity. Neither of these were in any sense commercial power producers: the honour for this lay with Britain's Calder Hall, a two-reactor station with a net design output of 70,000 kilowatts.

Let us look at the basic design problems and see how these were tackled and crystallised out into the Calder Hall design. Basically each of the Calder Hall reactors is like one of the Windscale reactors up-ended and put into a closed vessel connected top and bottom with four smaller closed vessels which act as the shells of water-tube boilers. The whole system is filled with compressed carbon dioxide gas which is circulated by powerful fans upwards through the reactor vessel and down through the boilers. Thence it returns into the reactor vessel. In this way heat from the fuel rods is transferred to the boilers where it raises steam to drive more-or-less conventional turbines and alternators; the steam is condensed and pumped back

Fig. 9   *The Calder Hall nuclear power station showing the four reactor buildings each with four heat exchangers, and the two turbine houses.*

into the lower part of the boiler tubes. Two fluids therefore circulate separately: compressed carbon dioxide up through the core and down through the boilers, and water up through the boilers and on as steam through the turbo-alternators and back through the condensers into the boiler (see figs 10 and 11).

There were other differences between the Windscale and Calder Hall reactors, especially in the fuel used. Windscale existed solely to produce plutonium, and the heat generated was a mere nuisance, so temperatures were kept down to a convenient level. The fuel — uranium metal canned in finned aluminium tubes — was not designed to stand up to the temperatures needed for raising steam that could drive turbines, so it had to be re-designed. Uranium metal, with some attention to its metallurgical qualities, could be made to resist temperatures well above 400°C but something new would be needed for the protective covering. An alloy was required which would stand up to 400 to 500°C in high pressure carbon dioxide gas while under irradiation by neutrons and gamma-rays, without deteriorating and without reacting or alloying itself with uranium metal. Lastly — and this was a property that had not concerned designers before, but which assumed almost supreme

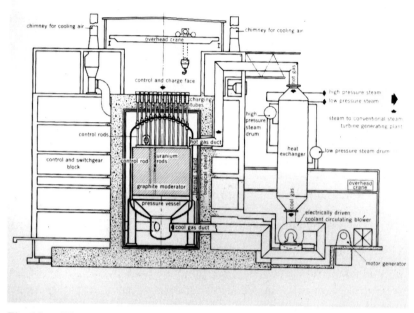

Fig. 10   *Diagram of one of the Calder Hall reactors.*

importance in the age of nuclear developments – it must not absorb
more than relatively small numbers of neutrons. After extensive
tests in BEPO under the conditions that would apply in Calder Hall,
an alloy of magnesium with small additions of beryllium, aluminium
and copper was chosen. Because of its high resistance to oxidation
in these conditions, this class of alloys was called 'Magnox' and has
given its name to the whole generation of power station reactors
based on the Calder Hall design and using magnox-clad uranium
metal fuel (see fig. 11). We consider the fuel in detail in Chapter 6.

   The Calder Hall design looked so promising that the original
twin-reactor station was duplicated and became a four-reactor
station while still at the drawing-board stage. Almost simultaneously
the Government authorised the building of a second, identical
four-reactor station at Chapelcross in Annan, just over the Scottish
border in Dumfriesshire. Both these power stations were to be
operated, at least initially, so as to produce the maximum amount
of bomb-grade plutonium, with electricity as a secondary product
– essential, but not to take precedence over the quality and
quantity of plutonium produced. This conflict of interests arose

because 'weapons-grade' plutonium is essentially the isotope Pu-239, but if this remains in the reactor for a long time it tends to absorb more neutrons and gives rise to higher and less effective plutonium isotopes, while non-fissionable radioactive decay products also begin to accumulate.

So for the most effective bomb material the fuel has to stay in the reactor for the shortest time consistent with worthwhile conversion of U-238 to Pu-239. If, however, electricity production is the primary aim, then the longer the fuel stays in the reactor the better: the more electricity it will produce and the lower will be its cost per kilowatt-hour. The limiting factors will be the progressive burning-up of the fissile uranium, the physical and mechanical endurance of the fuel, and the accumulation in it of neutron-absorbing fission products.

The first nuclear power programme

Meanwhile, events in the Middle East oil-producing countries made it clear that it would be sensible for Britain to look for an alternative energy source other than oil to supplement her coal resources. Even at the early construction stage it was clear that in the Calder Hall design we were on to a likely winner, and a programme of construction of twelve stations of the same general design, but slightly

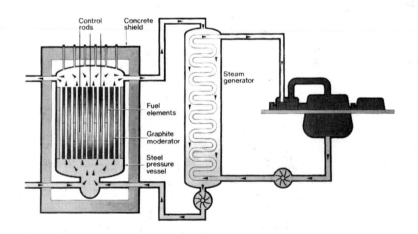

Fig. 11   *Schematic diagram of a basic gas-cooled reactor (MAGNOX).*

larger (50 to 100 MW electrical output) was announced in February 1955 by the Government of the day. Only a year later this was stepped right up to a programme that would give a combined electrical output of 5,000 MW upwards to be completed by 1965 (subsequently this was slowed down for economic reasons to a completion date of 1968) – see Table in Chapter 4. This was the first firm nuclear power programme adopted by any country in the world and it made a deep impression on all those attending the international 'Atoms for Peace' conference and exhibition organised by the United Nations in Geneva in 1955. Three years later at the 1958 Geneva Conference, with Calder Hall and Chapelcross in full operation and the commercial Magnox building programme well under way, Britain again stole much of the attention of the world. Let us look briefly at what some other countries were doing in the early post-war years.

Other countries

The United States had followed a very different line from Britain, although here too the link between military and peaceful developments was strong. Instead of deriving power-station reactor designs from plutonium-producing reactors as had been done in Britain, the Americans had, for strategic reasons, been looking closely at a very different type of reactor with a view to its use as a propulsion unit for nuclear submarines. This would have to be small in physical size but with a relatively large power output, capable of running continuously on a very small amount of fuel and with no need to 'breathe', as a diesel submarine must. Such a small reactor could only be made to work if the fuel was substantially enriched in its fissile U-235 content — a technique that was already well-developed in America for the U-235 bombs. This approach led to the development of the Pressurised Water Reactor (PWR) concept, and later to the basically similar though rather simpler Boiling Water Reactor (BWR). In both of these reactors water acts both as moderator to slow down the neutrons and as coolant to transfer heat to boilers and turbines (replacing respectively the graphite and carbon dioxide of Britain's gas-cooled reactors).

In the PWRs (see fig. 12) water circulates under high pressure through the reactor core to the boilers where it passes its heat to water-tubes in which steam is raised for the turbines. To prevent the water in the reactor circuit from boiling, and at the same time

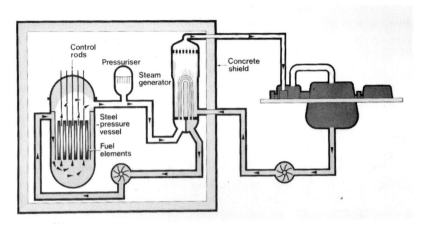

Fig. 12   *Schematic diagram of a Pressurised Water Reactor (PWR).*

to permit steam of good quality (i.e. high temperature and pressure) to be raised in the boilers, the reactor water circuit has to be kept under a pressure far higher than the carbon dioxide of Britain's Magnox reactors. Even so, the maximum temperature is limited by that at which water will remain liquid no matter how high the pressure, in practice about 280°C.

But water is in many ways not as good a moderator as graphite because it tends to absorb too many neutrons, and it can only be made to work with enriched uranium as fuel. This was, in any event, necessary if reactors were to be built small enough to use in submarines. Water offered the further advantage that it enabled uranium oxide to be used as fuel instead of the much less robust uranium metal used in the British reactors. In the USA there was already ample enrichment capability. If the water reactors were suitable for submarines, then scaled up to larger sizes they should be equally suitable for land-based power stations. However, enriched fuel would be more costly, and difficulties might be expected in the construction of power-station pressure vessels robust enough to stand the extremely high pressures anticipated.

These considerations were among those leading to the development, for land-based power stations, of an alternative to the PWR – the Boiling Water Reactor (BWR) (see fig. 13). Here the water is allowed to boil as it passes through the core, and the steam goes directly to the turbines from which it returns, condensed back to water, to the reactor core. This is a much simpler system, doing without the pressuriser and the heat exchanger of PWRs, but it has the drawback that the water itself becomes mildly radioactive as it passes through the core of the reactor, so special precautions become necessary in the turbine area of the power station, as well as the reactor itself. This imposes some constraints on access to the turbine house for maintenance. However, the two systems, PWR and BWR (collectively known as Light Water Reactors – LWRs), have been competing on more or less equal terms for the power-station market, not only among the public and private utility companies of the USA, but also for the export trade. Such has been the selling vigour of the major American companies, that the western world's market for nuclear power stations (other than in Britain and Canada) was for many years dominated by them and their licencees.

France, meanwhile, some of whose scientists had been working on the joint Anglo-French–Canadian atomic energy project during the war, started by following a similar line of approach to Britain. She built a number of large graphite-moderated gas-cooled reactors for producing plutonium, and a reprocessing plant, and carried out her first weapon test in February 1960. Still travelling the same path as Britain she built three dual-purpose power plants, similar in basic design to Britain's Magnox reactors. Initially these were optimised for the production of weapons-grade plutonium and

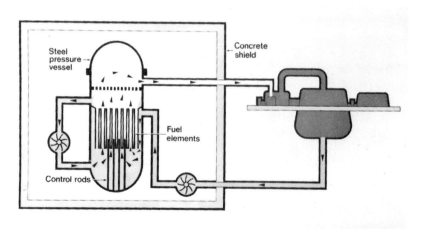

Fig. 13   *Schematic diagram of Boiling Water Reactor (BWR).*

later for electric power. From the mid-1960s, however, France built no more graphite-moderated gas-cooled reactors, but went over to pressurised water reactors for her further programme of nuclear power generation.

Canada had been the scene of an important part of the wartime effort on slow-neutron fission research. Much of the work there had been done by French and British as well as Canadian scientists, and there was some co-operation with the United States project. The working teams were therefore able to form a unit which could be regarded as an already viable project. It was in fact in the laboratories of Chalk River and Deep River in the province of Ontario that a small team of British atomic scientists under Dr John Cockcroft gained their experience and much of the know-how to set up Britain's establishment at Harwell, and to develop the production facilities so quickly.

In spite of the close links with the USA on the one hand and Britain on the other, Canadian post-war development went in a different direction from either. Renouncing military developments, Canada aimed directly for economic generation of electricity based on uranium from her own resources. The design that was eventually (and very successfully) adopted went by the felicitous name of CANDU (Canada Deuterium Uranium). It relied on natural (un-enriched) uranium oxide as fuel and aimed to extract the maximum amount of heat from it without considering the recovery of plutonium, or indeed, any reprocessing of the fuel. Neither water nor graphite would be good enough as a moderator and it would be necessary to use 'heavy water' — water in which the ordinary hydrogen isotope of mass 1 is replaced by 'heavy hydrogen' (deuterium) atoms of mass 2. This is also used to transfer heat to boilers where steam is raised to drive the turbines.

Heavy water is very expensive to make and forms a substantial part of the total investment in a CANDU reactor. To minimise the requirements and to reduce losses it operates in two separate circuits — one in pressure-tubes containing clusters of fuel elements, to carry heat from them to the boiler, the other in a tank or 'calandria' surrounding the pressure-tubes where it acts as a moderator for the neutrons. The heavy water in the calandria is kept cool, while that in the pressure-tubes operates at the highest practicable temperature and pressure (see fig. 14). The first CANDU power station started at Douglas Point, Ontario, in 1967 and was the forerunner of a large and successful programme.

Meanwhile Russia had not been idle. At the end of the war, al-

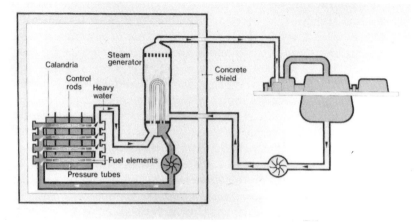

Fig. 14   *Schematic diagram of a CANDU heavy water reactor.*

though without the bomb — or indeed any major research and development effort — she lost no time in catching up. By September 1949 she had exploded her first nuclear test weapon, based on plutonium from graphite-moderated water-cooled reactors, and a chemical separation plant, and her first hydrogen bomb test followed in 1953. She soon looked to electric power production, initially from the same sort of reactor as those used for plutonium production but later from PWRs, which also formed the basis of her very large nuclear submarine fleet. Russia ventured early into another development that has proved very successful — the nuclear-powered ice-breaker (see Chapter 11).

The only other country to enter openly on a nuclear weapons programme was China who exploded her first test weapon in 1964. It seems likely that she will rely for civil nuclear power development largely on imported technology, at least in the early stages. India started early on civil nuclear research, particularly in the development and use of radio-isotope techniques in which she was greatly assisted by Canada who provided a research reactor. In 1974 she exploded a nuclear device ostensibly as part of a programme of peaceful research in civil engineering and allied work, following the American 'Plowshare' project for harnessing nuclear energy directly to such activities as canal construction and the exploitation of underground oil and gas reserves. India's action, however, was seen by many other countries as a clear and deliberate demonstration that she was in a position to use nuclear weapons, if necessary. At the time of writing no other country has yet given a similar demonstration of nuclear weapons capability, but there is little doubt that a number could do so, if they so wished.

Many other countries started in the 1960s and 1970s to develop civil nuclear power programmes after first cutting their teeth on small research and training reactors supplied under various international aid schemes, especially those sponsored by the International Atomic Energy Agency (IAEA) of the United Nations. Most countries eventually adopted PWRs or BWRs built by or under licence from the United States makers or by European firms which had developed their own variants. A few countries opted for CANDU reactors while Italy and Japan were alone in each buying a British power station of the Magnox design. The countries of the Soviet Bloc not surprisingly used Russian-based technology for their power programmes.

In most of those countries that have entered into programmes of nuclear power the proportion of nuclear to total electricity generation now averages roughly 10 per cent.

It is worth remembering that the world's very first commercial electric power station at Holborn in London was opened less than a century ago. It used coal as its fuel of course, but as the industry grew oil captured an increasing share of the electricity generation market, also water power (hydro-electric generation) in suitable regions. In its short life of under twenty-five years the nuclear industry, using uranium as its primary fuel, has thus already made a very sizeable impact on the electricity supply situation. It must be noted, however, that the rapid initial growth has in recent years been slowed down by the increasingly vocal public concern for the environment spearheaded by protest groups, of which Friends of the Earth Limited are perhaps the best known. Nuclear power has been singled out because of its close association with radioactivity and with nuclear weapons. The grounds and merits of this reaction against nuclear generation are discussed in later chapters.

4

NUCLEAR REACTORS

Thermal and fast reactors

For practical purposes there are two basic types of nuclear power reactor: 'fast' reactors which use fast-moving (unmoderated) neutrons to keep a chain reaction going in plutonium (previously created from U-238) and 'thermal' reactors which use slowed-down neutrons to keep the reaction going in naturally-occurring U-235. Thermal reactors form the basis of almost all the world's nuclear power generation at present, so we will deal with them first.

Thermal reactors have a moderator to slow down the neutrons so that enough of them react with the fissile but comparatively scarce U-235 atoms to keep the chain reaction going. Although many light atoms will in theory, slow down neutrons, the choice of moderator in practice is virtually limited to carbon (graphite) or water (including heavy water). This choice leads to two main families of thermal reactors: the first comprises graphite-moderated reactors and includes Magnox reactors (already described in Chapter 3) and Advanced Gas-Cooled Reactors (AGR), also mentioned briefly, and a range of experimental high-temperature reactors to be described later. In nearly all of these the heat generated in the reactor core is transferred to the boilers by compressed gas (usually carbon dioxide) circulating around the system from core to boiler and back.

The second thermal reactor family comprises the water-moderated reactors described in Chapter 3, p. 49. They include Pressurised Water Reactors (PWRs) and Boiling Water Reactors (BWRs) — collectively known as Light Water Reactors (LWRs). ('Light' water is of course ordinary water, H_2O, so-called to distinguish it from heavy water, D_2O.) This family also includes Britain's Steam-Generating Heavy Water Reactor (SGHWR). In all of these systems water, light or heavy, also transfers the heat. The group also includes CANDU (Canada Deuterium Uranium) which uses heavy water for both jobs.

Reactors and their functions

A nuclear reactor may be built for any of a number of different purposes. The purpose that most concerns us here is of course the production of useful energy for electricity generation (or, to a much lesser extent, for ship propulsion). Other purposes include the production of military plutonium, scientific research, the testing of nuclear materials, training in nuclear science and technology, and the development of new kinds of power reactor. Let us briefly consider some of them.

The production of plutonium for military purposes was, as we have seen, the original reason why nuclear reactors were built. If a plutonium bomb is to detonate effectively the plutonium must be very pure, both chemically (unadulterated plutonium metal) and isotopically, that is to say it must be almost all Pu-239 with as little as possible of the higher isotope Pu-240. The fuel containing the plutonium must be taken out of the reactor before it has had a chance to absorb more neutrons and form appreciable amounts of Pu-240. At the same time, in a reactor intended solely for plutonium production there is no need to consider the heat produced by the reactor except as an expensive nuisance to be disposed of by means of the cooling system — water in the Hanford reactors of the USA and air in Britain's reactors at Windscale and France's at Marcoule. It should be remembered that all nuclear reactors containing uranium produce plutonium: how much and how well this is done will depend on the size and design of the reactor and on how it is managed.

Just the opposite of this is needed in a reactor used to produce power: here as many of the neutrons as possible are used to keep the energy-producing chain reaction going, a high operating temperature is needed, the fuel stays in the reactor for as long as possible to give value for money, and burn-up some of the plutonium as it is formed in the fuel may be an advantage. Build-up of higher plutonium isotopes is no disadvantage. The Calder Hall reactors confirmed that either of the two rival requirements of plutonium and power production can be met in a single design, but they also confirmed that in operating the reactors a choice must be made on which to go for as the main product. In a dual-purpose reactor the value of the plutonium is a factor that may have to be considered in arriving at the cost of the electricity generated.

Research and Materials Testing Reactors are comparatively small reactors, designed not to produce power or plutonium, but as tools for scientific research and development, usually (though not exclusively) in the context of nuclear power. They may also be used as major items of equipment for training in nuclear science and technology. Their main job is to provide an abundant source of neutrons and the means of doing experiments with them. In this they correspond to the familiar Bunsen burner which provides a source of heat on the laboratory bench-top, while test tubes or crucibles provide the means of holding the specimens to be heated. In the case of the reactor the position is of course infinitely more complicated: unlike the Bunsen burner which is relatively innocuous, it is a copious source of damaging radiation, including neutrons, and several feet of shielding are needed to give protection to the operators. There is therefore a need for complex equipment to get the specimens in or out of the reactor core without letting out dangerous bursts of radiation. Means must also be provided for monitoring and controlling the environment of the specimen or system under test: for example, it may need to be held at a temperature of 2,000°C in helium gas at a pressure far higher than that of the reactor's interior (which may be at only 60°C and near to atmospheric pressure). When a specimen is taken out, it (and the rig that holds it) will be intensely radioactive and will have to be handled and examined by remote control through heavy shielding in the system of 'caves' or post irradiation examination (PIE) facilities associated with the reactor.

Materials Testing Reactors (MTRs) provide the means for small-scale testing (to destruction if necessary) of materials and components for use in power reactors, e.g. fuels, fuel cladding materials, instruments, etc. Because MTRs usually provide a very intense hailstorm of neutrons (they are often fuelled with highly enriched uranium), valid results of experiments can often be obtained in a very short time compared with the expected real lifetime in a power reactor.

Research reactors are also used for basic studies of how neutrons react with matter of all kinds, and here they are useful for scientific investigations right outside the field of atomic energy — in crystallography for example. Research reactors may also be equipped with facilities for the production of radioisotopes (radioactive chemicals and radiation sources) by neutron bombardment of appropriate chemicals, etc. in the reactor core (see Chapter 11).

In the UK two major Research and Materials Testing Reactors, DIDO (see fig. 15) and PLUTO, are in operation at Harwell, and the Atomic Energy Establishment at Winfrith in Dorset operate a

Fig. 15   *Interior view of the building housing the Research and Materials Testing Reactor DIDO at Harwell. The lead flask in position on the reactor top is for removing irradiated materials. The instrumentation is for controlling and recording the numerous experiments carried on within the reactor.*

number of specialised research reactors and assemblies, including ZEBRA for fast-reactor physics research. The DRAGON high temperature reactor was also sited at Winfrith but has since been closed (see p. 70). The experimental 100 MW(e) Steam Generating Heavy Water Reactor is still operational, also at Winfrith. There are also a number of small training reactors situated at or accessible to universities in the London area, the North Midlands and Scotland. More closely associated with the development of nuclear power are the experimental reactors built specifically to test the physics of proposed new reactor systems. In the earliest stages of development of a new system, 'exponential' or 'sub-critical' assemblies may be used which are not large enough to sustain a chain reaction on their own, but have to be fed with neutrons from an external source. The next stage may be thought of as a dummy or miniature reactor with many of the same design features of the reactor that it is leading up to — fuel, moderator, coolant, core layout, control system, etc. — but on a much smaller scale. It produces a minimum of power consistent with its operating at all, thereby cutting down the problems associated with heat disposal and radiation shielding and of course cost.

The final step on the way to a new design of power reactor is the prototype, where engineering comes into it as well as reactor physics. Components of the kind, if not always of the full size, of those to be used in the real thing are put to the test, as well as the behaviour of the reactor itself. A major function of the prototype is to reveal and iron out the snags in the system, whether they be in nuclear physics, engineering or materials, before incurring the cost and effort of full-scale construction. Experience has shown that it does not pay to skimp or leave out this step, or to go ahead with the final product on the evidence of too small a prototype. This was one cause of the difficulties met with in the earliest AGRs (see below).

Power reactors

A power reactor's job is to provide heat for raising steam to drive the turbines and alternators of an electric power station, so it must produce steam at the right temperature and pressure to do this efficiently. The laws of thermo-dynamics and the experience of engineers combine to call for the highest practicable steam temperature, in any case not less then about 250°C. There is little to be gained, however, by pushing up temperatures so high that special turbines are required to get the best results. The latest coal- and oil-fired power stations in the United Kingdom use a standard design of 660 megawatt turbo-alternators which require steam at about 540°C. This sets an upper limit, which is readily achieved by the AGR. The steam output must be immediately and exactly controllable all the time, both to meet changing power demands and to ensure safe operation. Reliability is essential, partly for obvious reasons that apply to any plant (such as maintaining power output and keeping costs down) and, more specifically, because of the difficulties of access to the radioactive parts of a nuclear installation after it has been in use: faults here can be very expensive and time-consuming to put right.

Because of the injurious nature of the radiations emitted in nuclear fission and from the fission products, and because of the very large store of potential energy in the fissile material in the reactor core (especially in power reactors), the need for safety must take precedence over everything else. It is discussed fully in Chapter 9.

Reactor control

In a nuclear system, the fuel is already in place — enough for many months' operation or even years — and it is held back from releasing its energy by the presence of enough neutron-absorbing control material to mop up any neutrons as they appear. When this restraint is removed the neutron population starts to grow spontaneously in a chain reaction, and heat release starts and grows with it. This growth continues until, by re-inserting some of the control material, the neutron population is stopped from growing further. If it is then kept constant with an average over the whole core of just one neutron per fission going on to cause one more fission, the heat output remains constant too. Similarly, to reduce power control material is inserted, and then partially withdrawn to stabilise the power at the new lower level.

There are other ways in which reactors can be controlled besides soaking up neutrons: the effectiveness of a water moderator can be varied either by altering the amount that is present, e.g. by raising the water-level in the core; or its composition can be changed by altering the proportion of heavy water to light water ('spectral shift'). Many other factors besides the control system itself influence the power output of a reactor. Some of these act immediately or in the short term, e.g. temperature changes, while others are more in the nature of a slow drift due to changes in the fuel or moderator. But in either case, they will require the attention of the operating staff. One immediate factor is power demand: if demand for electricity rises more steam is at once drawn from the boilers to keep the alternators running at their proper speed and this in turn draws more heat from the reactor and suitable control action has to be taken — control rods are momentarily withdrawn to let the chain reaction build up to greater strength. Conversely, a sudden drop in power demand, caused for example by a power transmission line being put out of action by lightning, immediately reduces the electrical load and hence the heat demand from the reactor. To avoid over-heating action has to be taken, this time to reduce the neutron population and hence the power output. If rapid enough action is not taken by the operator or by the automatic control system, built-in safety devices come into play to avoid a dangerous situation by shutting down the reactor if need be (see fig. 16).

In most reactors changes in temperature of the fuel or moderator automatically bring about certain changes in neutron behaviour which can reflect in quite complicated ways on reactor power output. In a graphite-moderated reactor the overall effect of a

Fig. 16 *Part of the control room in the CEGB's 1,320 megawatt Advanced Gas-Cooled Reactor at Hinkley Point 'B' power station.*

temperature rise is generally for the reaction rate to drop (negative coefficient) or at least to rise only slowly (positive coefficient). In any case the changes are very well within the control of the reactor operator. In water reactors the picture is even more complicated: if the temperature rise produces additional steam-bubbles the efficiency of the moderator drops but so does the ability of the coolant to remove heat from the fuel.

We have seen that as the fuel in a reactor is gradually burnt up and fission products accumulate, its effectiveness drops, particularly in the most active regions of the core. This can be evened up by 'shuffling' the fuel as described on p. 107 and replacing some of it, perhaps before it is strictly necessary. It can be mitigated by initially incorporating 'absorbers' in these regions: these are materials which will soak up neutrons fairly strongly and keep the reaction in check locally so as to even things out — which is an engineering advantage as well. The principle can be extended to use 'burnable poisons' which gradually lose their power to absorb neutrons in step with the drop in the supply or neutrons as the fuel is consumed; or moderator composition can be altered as described above. Let us look now at some specific examples of power reactors.

Magnox reactors

As described earlier, the UK Government in 1954 announced a nuclear power programme based on reactors similar to, but bigger and better than, Calder Hall. Eventually nine stations having a total output of 5,000 MW were authorised, to be in full operation in 1968. As was to be expected with such a novel technology not everything went according to plan, delays and cost overruns occurred, some design aims were not achieved and some faults developed. For example, operating temperatures of some reactors had to be reduced below the planned figure because of relatively minor corrosion troubles within the reactor core region. However, with few lapses, the stations have all worked very well and are now regarded as the 'reliable work-horses' of the electricity generating industry. They represent about 4 per cent of Britain's total generating capacity, but between them they produce at the time of writing (summer 1979) about 13 per cent of Britain's electricity (see Table, fig. 17). The reason for this difference lies partly in the operating characteristics of gas-cooled reactors, in particular their

Britain's nuclear power stations

MAGNOX stations	Date of commissioning	Nett capability
		MW sent out
Calder Hall*	1956	200
Chapelcross*	1958	200
Berkeley	1962	276
Bradwell	1962	250
Dungeness 'A'	1965	410
Hinkley Point 'A'	1965	430
Hunterston 'A'	1964	300
Oldbury on Severn	1967	416
Sizewell 'A'	1966	420
Trawsfynydd	1965	390
Wylfa	1971	840

AGR stations		Nominal capacity
		MW
Windscale*	1962	33
Hinkley Point 'B'	1976	1320
Hunterston 'B'	1976	1320
Dungeness 'B'	1980	1200
Hartlepool	1981	1320
Heysham 'A'	1981	1320
Heysham 'B'	Late 1980s	1320
Torness	Late 1980s	1320

Other stations		Gross capacity
		MW
SGHWR, Winfrith*	1967	100
PFR, Dounreay*	1975	250

Fig. 17 *In the Table, the reactors marked with an asterisk* were built as UKAEA experimental reactors. The others were built by British industry, on the basis of UKAEA designs, for the Generating Boards to supply electricity commercially to the National Grid.*

fuel, and partly in their economics. Continuous operation at a steady level, even at the highest output, is less demanding on the fuel and moderator than repeated changes in output, especially when these involve cooling and re-heating both the fuel itself and the enormous bulk of the graphite moderator. Economically too, the Magnox reactors are best operated continuously because, by comparison with coal- and oil-fired stations they are expensive to build but cheap to run. This economic effect has been enhanced in recent years: in most countries the overall cost (capital cost plus running costs) of nuclear-generated electricity has fallen well below that for coal or oil. The biggest demand — the continuing baseload — is met as far as possible from the cheapest source of supply, that is to say the nuclear stations: if one of these is out of action the demand has to be met from a more expensive source.

To make near-continuous operation feasible routine shutdowns must be kept as few and short as possible. Of all routine operations on the reactor refuelling is by far the most important and the commercial Magnox reactors are equipped with the means of doing this without shutting down or depressurising the reactor. A typical refuelling machine for a Magnox reactor consists essentially of a massive cylindrical shielding structure, which can be positioned on the reactor top shield immediately over any selected fuel channel. Here it is clamped firmly down so that there can be no escape either of the carbon dioxide coolant gas or of radiation. An internal gripping and hoisting mechanism removes first the shielding plug over the fuel channel and then the fuel elements, one at a time, using the end fittings to grip onto. The elements are stowed within the machine's shielding, where a water cooling system removes the very considerable heat generated by radioactive decay of the fission products. The machine then replaces the spent fuel with fresh fuel (or with partially spent fuel from another part of the reactor) and moves on either to another fuel element channel or to desposit its load in the spent fuel storage area (also shielded and cooled). From here the fuel will later be taken to the power station cooling-pond for storage prior to its removal to the reprocessing plant (see Chapter 7).

Advanced gas-cooled reactors

In 1964, when the Magnox reactor building programme was about half-way through, the British Government announced a second programme. There was much discussion about what design of reactor to use, with the field of choice including Magnox, the American light water reactors and Britain's Advanced Gas-Cooled Reactor (AGR), the last-named being a logical development of the

Magnox reactor principles, (to be dealt with in the discussion on nuclear fuel in Chapter 5). The choice was based partly on technical merit, but mainly on estimates of the comparative costs of electricity from each system as tendered for by the groups of firms in the business of nuclear power station construction. The lowest tender was for an AGR station. Experience with a small (33 MW(e)) prototype AGR at Windscale (WAGR) had already proved the technical merits of the fuel and confirmed the feasibility of the basic reactor design. Accordingly, in 1964, the Government authorised the building of five AGR stations, and construction was to start at once on the first of these at Dungeness on the Romney Marshes in Kent, alongside an existing Magnox station. It was to have two reactors each giving an electrical output of 600 MW(e).

The AGR design (see fig. 18) developed logically from the Magnox graphite-moderated gas-cooled family of reactors, but without the same limitations on steam temperature. The major changes, however, were in the fuel — to be described in Chapter 5. The fuel can operate at much higher temperatures and heat output rates than Magnox fuel, giving gas outlet temperatures and pressures of about 650°C and 600lb/sq.in. as against 410°C and 300lb in Magnox. This gives much better steam conditions in the turbines and leads to more efficient electricity generation: the best that Magnox stations can achieve is about 31 per cent steam cycle efficiency, while AGR stations aim for about 42 per cent — as good

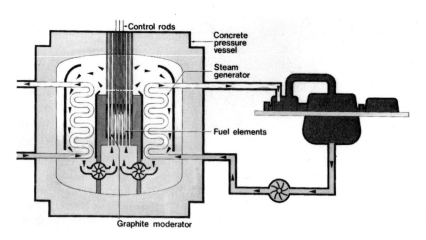

Fig. 18   *Schematic diagram of Advanced Gas-Cooled Reactor.*

as contemporary coal- or oil-fired stations. This opens up the possibility of using the standard design and size (660 MW(e)) of turbo-alternator used in the Generating Board's latest coal and oil stations. Another feature of the AGR design places the boilers inside the pressure vessel instead of outside it, as in all but the latest Magnox stations, thus shortening and simplifying the coolant gas circuit. Also, as in the later Magnox designs, the reactor vessel is of pre-stressed concrete built in one piece with the biological shielding, with a steel lining to protect the concrete.

The Windscale prototype had successfully pioneered many of the features planned for commercial AGR stations, so construction went ahead. But WAGR was only a 33 MW(e) reactor and those at Dungeness 'B' were to be nearly twenty times this size. The scaling-up was too big: design shortcomings soon became apparent and unexpected difficulties arose in construction and assembly, particularly in installing the boilers. There was heavy vibration from the gas blowers, parts of the heat insulation and some components of the internal fuel handling mechanism failed under the fierce blast of the coolant gas and had to be removed, re-designed and replaced. In addition to technical problems, the group building the reactor ran into grave industrial and financial difficulties. There were long delays and heavy overrunning of costs, and eventually the contract was taken over by another group, the newly-formed Nuclear Power Company Limited (NPC) which in 1975 became the sole builder of nuclear power stations in Britain's newly reorganised nuclear industry. (This rationalisation of the industry had taken place during the early 1970s, reducing the number of nuclear construction contractors from an original five groups at the start of the Magnox programme to three, then two, and finally in 1973 to one only – the National Nuclear Corporation (NNC), in which the UKAEA holds 35 per cent of the shares on behalf of the Government, and of which the Nuclear Power Company (NPC) is the operating arm.) Dungeness 'B' is now expected to come into operation in 1980 as a viable power station producing electricity at a cost competitive with coal. When considering troubles, delays and excess costs of power stations it is important to remember that these happen in coal-fired stations as well as in nuclear ones, but they tend to get less publicity.

Meanwhile the rest of the AGR programme has gone ahead, with NPC building four other stations of which Hinkley Point 'B' in Somerset (see fig. 19) and Hunterston 'B' in Strathclyde came into operation in 1976. The others, at Heysham in Lancashire and

Fig. 19   *General view of Hinkley Point power station in Somerset. On the left is the 550 megawatt Magnox station and on the right the 1,350 MW AGR station. Note how much more compact the latter is, in spite of its much higher electrical output.*

and Hartlepool, Cleveland, are due to start generating in 1981. They are all twin-reactor stations having electrical outputs of 1,320 MW(e) from two standard 660 MW turbo-alternators. With the Magnox stations, the five AGRs will bring nuclear power generation in Britain up to 20 per cent of the total electricity needs of the country, or about 5 per cent of total energy needs. Two further AGRs are to be built later: work on the first of these at Torness, near Dunbar in Scotland, has already been started.

Other graphite-moderated reactors

A third member of the family of graphite-moderated thermal reactors using gas as the coolant is the High Temperature Gas-Cooled Reactor (HTR), which is still at an early development stage. It can work at very much higher temperatures than even the AGR, and a

Fig. 20 *Map showing the principal nuclear installations in the United Kingdom.*

larger proportion of the uranium fuel content can be burned up in the reactor. The high operating temperatures — 700°C to 1,200°C — are achieved by doing away with the metal fuel cans and distributing the fuel more or less uniformly through the graphite moderator, rather than as large pieces separated by several inches of graphite. Typically, HTR fuel takes the form of enriched (6 per cent U-235) uranium oxide or carbide grains about 1 mm in diameter, coated with successive layers of silicon carbide and carbon, and embedded in graphite (see fig. 21). This is machined to suitable shapes for the particular reactor design — tubes, rods, rings, spheres, etc. — and, with further graphite moderator, is assembled to form the reactor core. Helium gas is used as the coolant because at these high temperatures carbon dioxide would react rapidly with the graphite. The high gas outlet temperature leads to the interesting possibility of using at least some of the heat to drive a gas turbine engine rather than to boil water and raise steam in the conventional manner. Another possibility held out by HTRs is to incorporate some

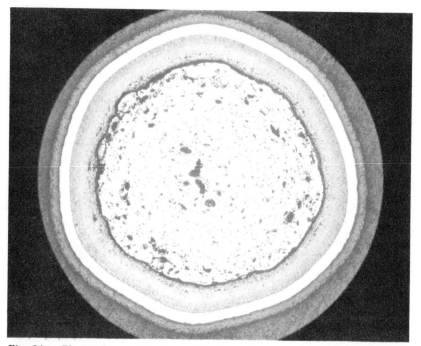

Fig. 21 *Photomicrograph of a polished section of a coated fuel particle as used in the experimental DRAGON reactor. Diameter about 1 mm.*

thorium in the fuel where it would be converted by neutron bombardment into the artificial fissile isotope uranium 233, analogously with the conversion of uranium 238 to fissile plutonium 239. This, however, is at an early stage of development and there are many difficulties to overcome before the world's reserves of thorium can be opened up for exploitation as a second nuclear fuel (see Chapter 12, p. 163).

The only HTR to be built in Britain was a small experimental reactor based on work started at Harwell as early as 1956. By 1959 a design had been prepared, and 15 European countries signed an agreement to build an experimental reactor at the UKAEA's establishment at Winfrith Heath in Dorset — the international DRAGON project. Construction started in the following year and the reactor came into use at its full power of 20 MW thermal (it did not produce electricity) in 1966. The reactor was used widely to test the physics and some of the engineering aspects of this type of reactor, to try out a range of different fuels and fuel element designs and layout (including thorium-containing fuels), to study fission product behaviour in the fuel, and to develop the use of helium gas as a coolant. The DRAGON project lasted for ten years to March 1976. Technically, it vindicated the belief that the HTR principles were sound and, with further development and scaling-up, could lead to a viable reactor system with potential for uses outside the power generating industry — process heat for steel-making, for example (see Chapter 11).

After the closure of the project in March 1976 work on HTRs ceased in Britain, but West Germany went on to develop a 'pebble-bed' high temperature reactor in which the fuel takes the form of a large number of spheres of about cricket-ball size; these move gradually downwards under their own weight through the active region of the core and out at the bottom, where they are automatically sorted on a basis of the burn-up each has achieved. Those good for further use are returned to the reactor, while the used-up ones go for reprocessing or storage. The United States are also working on their own designs of HTR, but commercial development of this family of reactors has not yet begun.

Graphite-moderated water-cooled reactors

Another type of graphite-moderated reactor uses pressurised water as the heat transfer medium. It was developed in the Soviet Union in direct line of descent from the early water-cooled plutonium-producing piles used both there and in America. Three such power stations formed the first instalment of Russia's power programme, but later, like so many other countries, she went over to PWRs.

Water-moderated reactors

We now turn to the reactors that use water as moderator. There are two groups of these: those using ordinary (light) water which include the PWRs and the BWRs of American origin, and those that use heavy water, comprising the Canadian CANDU, and our own Steam-Generating Heavy Water Reactor. These have already been described in Chapter 3.

SGHWR was conceived in order to provide experience with water reactors, and as a possible alternative to Britain's continued reliance on graphite-moderated gas-cooled reactors, but without following (like most countries) in the footsteps of the United States and choosing PWRs or BWRs. It has much more in common with CANDU than with any other system, being based on pressure tubes for fuel and coolant and a calandria (tank) for the heavy water moderator (see fig. 22). The coolant is ordinary water which

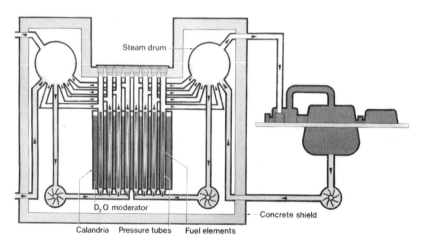

Fig. 22   *Schematic diagram of Steam-Generating Heavy Water Reactor.*

flows up the pressure tubes containing the fuel element clusters, where it boils. Steam from the top of the pressure tubes collects in steam-drums from which it passes to the turbines. The separated water, together with that condensed from the turbines, goes back to the bottom of the pressure tubes for re-circulation. The fuel consists of slightly enriched (2.6 per cent U-235) uranium dioxide canned in zirconium. In 1968 a 100 MW(e) SGHWR was commissioned at the Authority's Winfrith establishment where it remains in successful operation as a small and highly reliable power station and as a test facility for water-reactor components. Efforts to sell this design overseas were not successful and the Government's decision in 1974 to base Britain's third nuclear power programme on SGHWR was revoked in 1978 in favour of more AGRs, to be accompanied by design work on a PWR for later ordering. Development costs proved the main obstruction to progress and SGHWR still has much to be said for it.

Further advances in thermal reactors

The stage has been reached in thermal reactor research when the lines of advance have become fairly well established, and the future now lies less in new ideas than in detailed improvements of existing well-tried designs. For example, the introduction of a greater measure of standardisation would cut design and manufacturing costs, speed up construction and reduce the number of variables to be considered by a prospective buyer of a reactor system. Concentration on safety factors and well-proven designs is likely to increase public confidence in nuclear power generally. Another likely line of advance lies in the development of reactors of established types in much smaller units but without losing the economic advantages that large reactors have: not every potential reactor user wants an output of a thousand or more megawatts and there may well be a market for units in the 20 to 200 MW range — provided that they will produce cheap power. Improved load-following — the ability to respond well to continued changes in power demand — is another field where improvements can be made.

The most promising area for major technical advances appears to lie in the development of high-temperature gas-cooled reactors to be used in association with gas turbines or chemical process heat, and possibly with the use of thorium as a fuel (see also Chapter 12). Development of heavy water reactors — CANDU in particular — to burn thorium looks to be another possible line of advance.

Fast reactors

As we have already seen, it was known from the early days of nuclear fission research that uranium 235 could be split by neutrons moving at any speed. However, if U-235 was very much diluted by U-238, as in natural uranium, then to keep a chain reaction going, it was essential to use slow neutrons, because at high energies too many neutrons are captured by U-238 and wasted. In fact, experiments and calculations showed that if the fuel was rich enough in U-235 there would be no difficulty in getting a chain reaction going with fast neutrons only and no moderator. Further studies confirmed that plutonium too could be used as a fuel for a fast-neutron chain reaction, and that a plutonium-fuelled fast-neutron reactor would show very good neutron economy, providing a clear surplus of neutrons over and above those needed to keep the chain reaction going. Some of these neutrons could be absorbed by uranium 238 nuclei which would then turn into plutonium. Under the right conditions new plutonium atoms would be formed at a higher rate than existing plutonium was used up, and this could be extracted and used for fresh fuel. In other words, it might be possible for a fast reactor fuelled with plutonium, and kept supplied with uranium 238, to produce more fuel than it consumed — to 'breed' in fact. In this way the plentiful isotope uranium 238 might be regarded as a source of fuel (but not directly as a fuel itself) alongside the rare isotope uranium 235. The amount of energy available from raw uranium would thus be multiplied by a very large factor — 50 to 60-fold is a reasonable estimate based on experience up to the present time. But the term 'Fast Breeder Reactor', though widely used, is misleading: to breed enough plutonium to fuel a similar reactor would take about twenty years — hardly fast breeding.

Britain's fast reactor programme

From 1951 the potential of fast reactors was recognised in Britain and by 1953 — before Calder Hall was built — a very small fast reactor, ZEPHYR (the letter Z stands for zero energy), was in

Fig. 23   *Aerial view of Dounreay showing Prototype Fast Reactor in the fore-ground and the sphere housing the Dounreay Fast Reactor (now closed down).*

operation at Harwell. It confirmed the feasibility of the fast reactor concept (a conclusion which was supported by parallel work in the USA and Russia) and it led to the preparation of designs for a fast reactor large enough to operate a small power station. Building was authorised in 1954 and by 1960 the Dounreay Fast Reactor (DFR) was in operation in the north of Scotland, feeding 15 MW of power into the grid. It continued to operate successfully, both as an experimental and testing reactor and as a power station, until 1976 when it was closed down, being superseded by the much larger Prototype Fast Reactor (PFR), designed to generate 250 MW (see fig. 23).

PFR was designed on the basis of experience with DFR, which had provided an essential test-bed for PFR reactor components and materials, in particular for the fuel. At the same time, the problems of fast reactor physics were ironed out in the specially built research reactor, ZEBRA, at Winfrith. DFR had started with a core of highly enriched metallic uranium as fuel (75 per cent

U-235) canned in niobium, but as plutonium became available the fuel was changed to plutonium metal mixed with uranium canned in stainless steel. The core was surrounded by a blanket of metallic uranium 238 rods. PFR uses a core of a more fully developed fuel consisting of mixed oxides of plutonium (20 per cent) and depleted uranium (80 per cent) canned in stainless steel, with a blanket of depleted uranium oxide, also canned in stainless steel (see fig. 24). PFR fuel closely resembles the fuel that is likely to be used in a commercial fast reactor, and the same applies to reactor components such as boilers and sodium pumps. Because of the absence of any bulky moderator, a major problem in fast reactor design concerns the high rate of heat transfer from the core: DFR, for example, had a core of only the same size as a domestic dustbin, but in operation it produced up to 60 megawatts of heat — the same as 60,000 single-bar electric fires. (In contrast, each of the Calder Hall reactors produce about two and a half times that amount of heat, but it comes from a core about 35 feet across.) In DFR the only practicable way to get the heat to the boilers fast

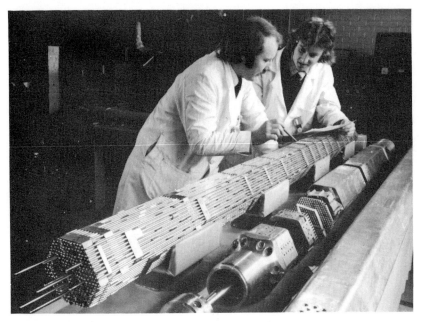

Fig. 24 *Prototype fast reactor fuel sub-assembly showing stainless steel fuel pins with wrapper and other components.*

Fig. 25   *A dummy fuel element being loaded into the core of the Prototype Fast Reactor at Dounreay.*

enough was to use a molten metal coolant consisting of sodium-potassium alloy (or pure sodium in PFR). A whole new technology of heat transfer by molten metal had to be developed, which came to involve handling the liquid metal in tonnage quantities:

it turned out to be surprisingly simple provided that it was kept scrupulously clean and free from oxide or hydroxide. In fact, sodium has proved to be a much less corrosive and troublesome material than water under normal boiler conditions. To minimise the consequence of a possible leak between sodium and boiler-water the reactor fuel gives its heat to a primary circuit of sodium which passes it on, via an intermediate heat exchanger, to a secondary sodium circuit which heats the boilers. In this way there is no risk of radioactive sodium from the primary reactor circuit reacting with boiler-water and possibly reaching the environment (see fig. 26).

Fast reactor fuels contain plutonium, and the technology of their manufacture and reprocessing had to be developed from scratch. Lengthy endurance tests had to be carried out on the fuel, using DFR as a test-bed to simulate the conditions to be encountered in the larger PFR which, in turn, simulates the conditions in a possible full-scale reactor. Structural materials have to stand up to reactor conditions for the whole lifetime of the plant, and here the particle accelerators at Harwell have proved invaluable for irradiation tests simulating the effects of prolonged fast reactor irradiation on such qualities as toughness of stainless steels.

Many of these problems have now been solved, or at least the solutions are in sight, and several countries, notably Britain, France,

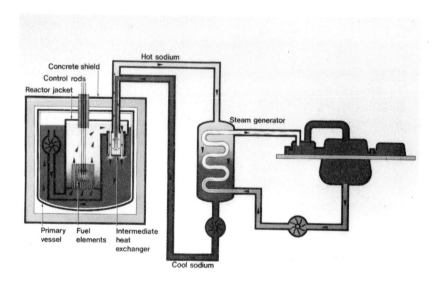

Fig. 26   *Schematic diagram of Sodium-Cooled Fast Reactor (PFR).*

Germany, USA and the Soviet Union, have fast reactors of broadly similar kinds at or near the stage of large prototypes of 200 to 600 MW(e). However, it is likely to be a number of years before fast reactor construction goes ahead on anything like the scale of thermal reactors at the present day. Other developments in fast reactors, for example gas-cooled fast reactors, show some promise but they lie a long way ahead.

Fast reactor development

The development of fast reactors has been slow compared with that of thermal reactors: at least five distinct kinds of thermal reactor have been developed to the stage of industrial use since the early 1950s, whereas in almost the same period only the one kind of fast reactor has been developed and at present this has not got beyond the large prototype stage anywhere in the world.

At the time of writing it is not possible to forecast when Britain will go ahead to the stage of building a fast reactor of full commercial size — probably 1,000 to 1,300 MW(e) — as a demonstration of feasibility. CDFR (Commercial Demonstration Fast Reactor) would be a logical forerunner to any possible future programme of fast reactors in the United Kingdom. It would demonstrate the feasibility of the design at full scale, reveal and help to iron out any remaining technical difficulties and give operational experience in commercial power production from fast reactors. All that can be firmly said at present is that in 1979 the Government renewed the promise to hold a full public enquiry to consider not only the merits of CDFR itself, but also the broad implications of a decision to build it.

5

URANIUM, THE FOUNDATION OF NUCLEAR POWER

Uranium as a mineral

Uranium is the basic fuel of nuclear power and it is from the world's reserves of this one element that man appears likely to be drawing a large and growing fraction of his needs for electric power for a very long time. Uranium is not very plentiful in the earth's crust, being rarer than silver but much less rare than gold. It is very widely distributed at low concentrations ranging from about one part per million (1 gram per tonne) in sedimentary rocks to about four times this figure in acid igneous rocks such as granite. Uranium is also present in sea-water at about one part per million in the dissolved solids. To be worth mining (at least for the present), however, much richer ores are needed and these occur in a wide variety of geological formations and in many parts of the world. Bulk deposits of low or medium grade ores (0.1 to 4 per cent uranium oxide) are commonest in or near to the very old pre-Cambrian rocks, or in conglomerates or sandstones closely associated with them. A few much richer deposits containing up to 15 per cent, and even more, of uranium oxide occur in veins widely distributed throughout the world.

Uranium, like gold, is where you find it, but before starting to search, it helps to know how it came to be there in the first place. There seem to have been two major ways in which workable concentrations have built up. In the first, molten rock from deep in the earth has cooled and selective crystallisation has taken place, leading to local vein-type concentrations of uranium (along with ores of other metals), mostly lying along faults and lines of weakness in the surrounding rock. These veins of comparatively rich ore occur in many different parts of the world and in many types of rock, but they only account for a small fraction of the world's uranium reserves.

In the second and more important process the original uranium-bearing rocks (granites, etc.), or regions where selective concentrations have already been achieved, are denuded and weathered by rain or attacked by hot underground water. The dissolved uranium finds its way along fissures or through porous rocks until it encounters chemical conditions which cause it to be thrown out of solution, either as a rich vein deposit or in a more dispersed form. This is likely to happen where the rock is basic rather than acidic, or where it contains chemical reducing agents or materials of organic origin such as carbon or hydrogen sulphide.

Pitchblende, in which uranium was first identified, proved to be a fairly pure oxide. Uranium was later found in a wide variety of ores, many of them mixed oxides or phosphates containing comparatively little uranium, along with other elements.

After the discovery of radioactivity in 1897 and the isolation of radium and polonium from pitchblende a few years later, uranium ores were mined vigorously for their medically valuable radium content, first in the Erzgebirge Mountains of Bohemia (Czechoslovakia) and later in Colorado. By 1915 major production had been moved to Portugal, but soon after the war ended, the very rich pitchblende mines of Shinkolobwe in the Belgian Congo (Zaire) took over as the world's main source of radium. They in turn ceased to be economical in the 1930s, when the mines of the Great Bear Lake in Canada were opened up. At around this time the market price of radium reached a peak of £8 a milligram.

With the discovery of nuclear fission in 1938 — almost exactly 150 years after Klaproth first recognised uranium — the position changed dramatically. Hitherto uranium had been a waste product of the mining of radium, but almost overnight it looked like becoming one of the world's most important strategic materials. Governments acted quickly to take over existing uranium supplies and to open up new ones: the tip-heaps of Shinkolobwe in particular became a military stock-pile of immense strategic value. Later large amounts of waste, as well as newly-mined ore, were shipped across the Atlantic to provide, with other sources, the basic raw material for the atomic bomb project. After the war the world-wide search for uranium began in earnest and, as with most hunts, the harder people looked the more they found.

At present the western world's major uranium-producing areas are in the United States, Canada, Southern and Western Africa (including Namibia), Western Europe and Australia. The largest known workable deposits are in the quartz conglomerates of

Elliot Lake in Canada, and in the Witwatersrand in South Africa. Here uranium is associated with gold and is to be found in workable amounts in the huge waste-tips of the Rand gold-mines. The USA has the largest sandstone deposits, in the Colorado Plateau, the Wyoming Basin and the Texas Gulf coastal areas. Vein deposits provide a smaller but much more widespread source. Sweden has enormous reserves of very low grade uranium-bearing shales. Little detailed information is obtainable about uranium resources available to Russia and the other Warsaw Pact countries or to China, but there is no reason to suppose that their position is significantly different from that of the West.

Finding uranium

Uranium is itself radioactive (though only very mildly so) and it is always accompanied in nature by the daughter-products of its own radioactive decay, and some of these, in particular radium and radon (a gas), give off easily detectable radiations. So the presence anywhere of radioactivity above the normal background level for the region may be an indication that here, possibly, is an increased concentration of uranium. But setting out in unknown country wearing a stout pair of boots and carrying a Geiger counter is not the best way of finding a new deposit of uranium: in the unlikely event of an enhanced level of radiation being detected, it is quite likely to be found to come from thorium and its daughter-products, or even from high levels of potassium, and by itself is no indication of the presence of uranium. Confirmatory tests are always necessary. In prospecting for uranium (as for any other element) the first need is for a sound knowledge of the geology and mineralogy of the element, supplemented by as good a prior knowledge as possible of the geology and topology of the area to be investigated. It is only after all the known facts about an area have been studied that decisions can be made on just where to concentrate the search, and what methods to use. It is not until this stage has been reached that radiation surveying comes into its own.

   If the area to be surveyed is large and hard to reach or to travel through, an air survey will usually be best. Typically the plane will fly at about 50–100 metres above the ground along grid-lines 150– 450 metres apart, logging radiation levels and keeping constant checks on position and altitude. Over some types of ground a simple Geiger counter may suffice as the detector, but it will usually be worth using the dearer but much more sensitive and effective scintillation counter, equipped with a 'gamma-ray spectrometer' to differentiate between radiations from the different elements (see fig. 27). As much additional information as possible should be obtained during the flight, and the survey plane may also carry magnetic and other geophysical instruments, with a cine camera to check location and keep a record of the topography.

Fig. 27   *Using a portable four-channel gamma-ray spectrometer for uranium prospecting.*

Most important of all, an experienced geologist familiar with the region should fly with the plane.

If ground conditions, time and the size of the area allow it, a much less expensive survey can be carried out using a Land Rover or other field vehicle carrying a detector mounted as high as possible from the ground: this gives the added advantage that interesting-looking outcrops or areas of high radiation can be examined on the spot and rock samples taken for further analysis.

Finally, detailed ground examination, with sampling of ores and the surrounding rock, will be needed in order to pinpoint the best places to make trial drillings. In areas where there is a thick surface layer of peat, for example, it may be possible almost literally to 'sniff out' underlying uranium by probing for a metre or so into the peat and testing the ground gases for the radioactive decay-product, radon.

Once a uranium deposit has been found and confirmed on the spot by radioactive methods supplemented by simple chemical tests, representative samples must be taken from the surface, from trenches and from boreholes, and sent back to the laboratory for detailed examination. Here advanced radiometric, chemical and other tests will be carried out on these and on samples from other locations to confirm and compare the types and grades of ore. A body of information is thus built up on which to base decisions on whether and where to start mining operations. It may take ten years or more from a decision to make a first exploratory survey flight to the arrival at the pit-head of the first batches of uranium ore.

Uranium mining

Uranium is mined in the same way as other metal-bearing ores. For example, round the edge of the granite area of the Great Canadian Shield almost all uranium is mined deep underground, while in the comparatively shallow conglomerates of South Africa and in the Colorado sandstones open-cast methods are more usual. The conglomerates around Johannesburg have already been mined for gold by both methods, and uranium is now being recovered as a by-product from the refinery waste-tips. Because of their origin and mode of formation many uranium deposits occur in rough country or in areas where the climate is particularly severe, and this can add substantially to the cost of mining and transport.

A drawback that applies particularly to uranium mining — though it can occur in other metalliferous mines as well — is the radiation arising from radioactive decay of the daughter-products of uranium. In particular, radium decays to the short-lived radon gas which, on inhalation, can deposit its own much longer-lived daughter-products in solid form in the lungs. Here they emit alpha-particles and may lead to the development of lung cancer later in life. (The tendency to an illness known as *Bergkrankheit* had been noticed in some pitchblende mines long before its nature was understood.) Similar trouble has also occurred in other metalliferous mines in areas where small amounts of uranium or thorium are present in the ore or the country rock. The danger, once recognised, can usually be overcome quite simply by ensuring adequate ventilation, damping down any dust and properly disposing of mine wastes and refinery tailings.

Because of its low metal content uranium ore is usually upgraded at the mine to a more concentrated form — 'yellow-cake' — containing 50 per cent or more of uranium oxide. The methods of wet chemistry are used rather than the hot smelting processes common with most other metalliferous ores. After initial mechanical crushing and sorting, in which rock lumps showing only low radioactivity may be rejected automatically, the selected ore is ground to a fine powder in conventional rod or ball mills preparatory to dissolving. At these stages quite large amounts of radon gas may be given off and precautions have to be taken against its inhalation and in the management of waste. Uranium itself is only mildly radioactive, being an alpha-emitter with a very long half-life of 4,500 million years. Its chemical toxicity is akin to that of lead and other heavy metals and, from the point of view of safety, attention to this fully covers any radioactive hazard the uranium may offer.

Handling, storing and transporting purified uranium and its compounds therefore present no special problem (although there is a very slow build-up of radioactivity from the newly-formed daughter-products arising from the slow decay of uranium itself). It is not until uranium has been subjected to the fission process in a nuclear reactor, and fission products have been formed within it, that significant radiological problems again arise.

The next stage in the process is to get the uranium into solution by a 'leaching' process, in which the crushed ore is treated with nitric or sulphuric acid, followed by filtration from the undissolved residues containing iron, copper, thorium and some of the valuable 'rare-earth' elements. Recent developments include alkali leaching and bacterial treatment to make the uranium easier to dissolve, while experiments are in hand on leaching the ore-bearing rock *in situ* without mining it.

Many other materials besides uranium find their way into the acid solution and the bulk of these are removed at the next stage. In the early days the solution was neutralised and the uranium directly precipitated by adding enough alkali to form an insoluble diuranate salt which was filtered off from the impurities remaining in the solution. But separation was by no means complete and the process of solution, precipitation and filtration had to be repeated to get a reasonably pure diuranate without too much loss of uranium.

The introduction of 'ion exchange' techniques, operating on similar principles to domestic water softening, simplified and improved the process considerably: the impure uranium solution with its acidity carefully adjusted passes through a bed of special synthetic resin particles having a strong selective chemical affinity for uranium ions. Like the dissolved calcium or magnesium entering a domestic water softener, the uranium forms a loose compound with the resin surface. Most other dissolved substances remain in the water, in which they leave the plant. When the bed of resin particles can take up no more uranium, the flow is stopped, and water adjusted to a different acidity is passed through the bed: the uranium is re-dissolved from the resin surface and can be reprecipitated with alkali as a relatively pure diuranate salt, and the bed is then ready for re-use. Ion exchange plants are usually built in duplicate, with one bed absorbing uranium while the other is being washed out.

Continuous solvent extraction techniques (already developed for the later stages of uranium purification) are also used in this part of the process. They depend upon uranyl nitrate being soluble not only in water but also in certain organic solvents. By carefully adjusting the acidity and temperature in a vessel containing water and solvent stirred together, the dissolved uranium can be made to go into the solvent while most of the associated impurities remain in the water. If now this water is run off and replaced by clean water with its acidity and temperature suitably re-adjusted, the uranium will leave the solvent and go back into the water. The system can be operated on a basis of continuous flow and counter-flow through a number of stages, with the acid being recovered for re-use and the same solvent being re-circulated. The impurities recovered from the original acid solution include a further range of by-product materials such as thorium, radium, vanadium, tungsten and rare-earth elements such as neodymium. The uranium is precipitated from the water solution by alkali and shipped as 'yellow-cake' of guaranteed uranium content.

Uranium purification

If a nuclear reactor is to work successfully, unnecessary neutron losses must be kept to a minimum. The fuel in particular must be free from those kinds of impurities that are strong absorbers of neutrons. As little as 1 gram of, for example, cadmium, boron or some of the rare-earth elements in a tonne of uranium may be too much. Achieving the required levels of purity has called for the introduction of standards of quality-control far and away above anything previously achieved in large-scale industry. To set and maintain these standards calls for careful organisation, high standards of housekeeping and working discipline, and the development of specialised analytical techniques applied at every stage to raw materials, chemical reagents and intermediate and finished products. In these respects the atomic energy industry sets an example which could well be followed by many other industries.

The first of Britain's atomic factories to be established was at Springfields, near Preston in Lancashire, where there had previously been a chemical factory run by the Ministry of Supply. In 1948 Springfields started supplying fuel for the plutonium-producing reactors at Windscale, also for the research reactor BEPO at Harwell. By 1954, when the whole of Britain's atomic energy facilities and staff were transferred to the newly-formed United Kingdom Atomic Energy Authority (UKAEA), the works was also supplying fuel for the Calder Hall and Chapelcross nuclear power stations, and was beginning to prepare the initial fuel charges for the earliest commercial Magnox reactors of the first nuclear power programme. In 1972 the Springfields works, together with the other production establishments of the UKAEA and the associated Trading Fund, was handed over to a new organisation, British Nuclear Fuels Limited (BNFL). This was, and continues to be, run as a commercial, profit-making organisation wholly owned by the Government with the UKAEA holding all the shares on behalf of the Government.

Springfields now has four major jobs: first, to convert the imported yellow-cake to a standardised uranium tetrafluoride powder of very high purity; second, to prepare to the user's design and specification nuclear fuel elements for Magnox reactors (metal fuel) and for Advanced Gas-Cooled Reactors or Light Water Reactors (enriched, oxide fuel); third, to convert uranium tetrafluoride powder into uranium hexafluoride for enrichment (see below) at Capenhurst or for overseas customers; and fourth, to reconvert enriched hexafluoride into oxide.

To give an idea of the scale of operations at Springfields, the chemical plant produces about 5,000 tonnes a year of uranium tetrafluoride powder, about half of which goes to feed the Magnox fuel production line. Most of the rest is converted to hexafluoride for enrichment and then re-converted to feed the enriched oxide fuel production lines. The fuel dispatched from Springfields every year to be burned in nuclear power stations is equivalent in, electricity production to about 14 million tons of coal or 8 million tons of oil.

All the chemical processes at Springfields, except for the re-duction of tetrafluoride to metal, are operated continuously and largely under automatic control. After the raw material has been sampled and checked for quality-control and record purposes, it is dissolved in a fairly strong hot nitric acid and the filtered solution of uranyl nitrate is fed into the first stage of the solvent extraction plant. This consists of a series of inter-connected 'mixer-settler' units, in each of which the acid solution is agitated in the 'mixer' section with an organic solvent and then allowed to separate out from it in the 'settler' section (see fig. 28). A dozen or so similar units are connected in series, with the solvent flowing in one direction and the acid solution in the other. At each stage the uranium tends to go with the solvent (a mixture of tributyl-phosphate and kerosene), while the impurities tend to stay with the acid. The two liquids flow continuously in counter-current fashion through a number of similar stages, leaving the uranyl nitrate solution with only a few parts per million of significant impurities. Pure uranyl nitrate is washed back into less acidified water in a second part of the plant similar to the first. The solvent is con-tinuously recycled through the plant, almost all of which is constructed from stainless steel — an expensive material requiring special fabrication techniques but essential to prevent the uranium solution from picking up impurities.

Purification is now complete and the uranyl nitrate solution, freed from traces of the solvent, passes to the next stage of the process where it is converted through a sequence of dry processes to uranium tetrafluoride. First it is concentrated by evaporation and sprayed into a reaction vessel in which an electrically-heated

Fig. 28   *The nuclear fuel works at Springfields: the uranium purification plant showing the mixer-settler lines. (Courtesy of BNFL.)*

bed of about 10 tonnes of yellow uranium trioxide (UO_3) powder is kept continuously agitated ('fluidised') by blowing carbon dioxide gas through it from below. This drives off the water, and breaks the uranyl nitrate down to form more uranium trioxide powder, giving off nitrous fumes which are recovered for re-use as nitric acid in the dissolving stage. As the uranium trioxide powder is formed it overflows continuously into a second and hotter fluidised bed, through which hydrogen passes. The incoming trioxide is reduced by the hydrogen to black uranium dioxide (UO_2) which in turn flows on to a third bed. This is fluidised by hydrogen fluoride gas (pure hydrofluoric acid) which converts the oxide to the green tetrafluoride (UF_4). Uranium tetrafluoride is the starting material both for uranium metal and for uranium hexafluoride (UF_6 or 'hex').

Metal production

Up to this stage the process is continuous, but for metal production a batch process is used. The tetrafluoride is mixed with shredded magnesium metal and compressed into cakes weighing about 3 kg each; these are stacked in graphite-lined vessels purged of air with the inert gas argon, and heated electrically. At about red heat a strong chemical reaction sets in and the temperature rises sharply. The fluorine atoms leave the uranium for the magnesium and a molten pool of uranium metal collects in the graphite catch-pot beneath a layer of magnesium fluoride slag. After cooling, the resulting billet of about 400 kg of pure uranium is removed and cleaned; it is sampled for analysis, numbered serially and stored in readiness for casting into fuel rods.

Enrichment

All oxide fuels (except those for the CANDU reactors) have to be up-graded ('enriched') in fissile uranium 235, otherwise the additional oxygen atoms would prevent the chain reaction from taking place. Moreover, the use of enriched fuel increases the amount of useful energy that can be extracted from a given quantity of raw uranium.

Isotopic enrichment is unlike any other industrial process, in that it involves separating two materials that are chemically identical and differ only in the masses of their atoms. Until recently the only practical method of separation has used the difference in diffusion rates of gases, which are inversely proportional to the square roots of their densities: a light gas will diffuse through a small hole (or through a lot of small holes grouped together to form a porous membrane) faster than a heavier gas.

Uranium itself is not a gas, but to be enriched it must first be turned into one. Fortunately one of its compounds, uranium hexafluoride or 'hex', forms a vapour at around 60°C. The weights of the two kinds of hex molecule, 343 units for U-235 and 346 units for U-238, differ by under one per cent and their diffusion rates by the square roots of these figures, that is, by about one part in 10,000. Only by using this minute difference over and over again can a useful separation be effected.

Uranium hexafluoride is made by adding two more atoms of fluorine to each molecule of uranium tetrafluoride. In the United Kingdom the starting material may be 'natural' uranium tetrafluoride from the main Springfields fuel production plant, in which case it will contain 0.73 per cent of the 235 isotope, or it may be 'depleted' material from the Windscale reprocessing plant (see below) which may contain as little as 0.4 per cent of U-235. The hex plant at Springfields, like most of the other plant there, operates continuously using a fluidised bed process in which tetrafluoride powder is fluidised by a current of fluorine gas. The two materials react together very vigorously (the reaction vessel has to be kept cool artificially) and the hex is formed as a vapour which is condensed directly as a solid in steel vessels designed to be used both for storage and for transport. The fluorine used in the process is prepared on the spot by electrolysis of magnesium fluoride slag from the metal production process — another example of the recycling of 'waste' products.

The isotope separation plant, originally commissioned in 1952, is operated by BNFL at Capenhurst in Cheshire. It consists of a very large number of identical units in a range of sizes. The largest units are where the hex is fed in and the size tapers off in both directions down to relatively small units at the output end. Each unit consists essentially of a rotary gas compressor (and associated cooler) feeding hex vapour into a chamber bounded on one side by a porous membrane. Most of the vapour leaves through an outlet pipe, but not before some of it has had time to diffuse through the pores of the membrane. Because the lighter molecules diffuse faster than the heavier ones the hex that gets

Fig. 29   *The nuclear fuel works at Springfields: the control panel of the 'hex' plant. (Courtesy of BNFL.)*

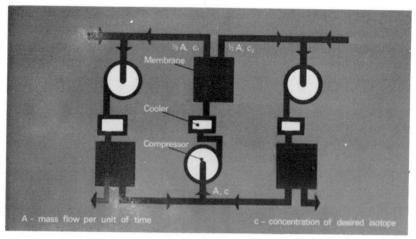

Fig. 30 *The cascade system of units in the uranium isotope separation plant at Capenhurst. (Courtesy of BNFL.)*

through the membrane will be just a little richer in the lighter isotope than the hex remaining in the chamber. The light (enriched) fraction passes on to the next unit up the line, where it is again compressed and cooled, and the process repeats itself. Meanwhile, the heavier (depleted) fraction that did not go through the membrane of the first unit leaves the chamber by the outlet pipe and goes on to feed the next unit down the line. Here a little more of the light component goes through the membrane and passes back up the line to form part of the feed to the first unit, and so on.

The number of stages in the whole system will determine how rich in uranium 235 the product can be made, while the total area of membrane will determine the output of the plant for any particular degree of enrichment. Since the amount of separation achieved in each single stage is very small, a great many units are required to make a successful separation. It is not practicable to build units in more than a few sizes, so hundreds, in some cases even thousands, of units are connected together in a 'cascade' system to give the required output of the whole diffusion plant. This may involve many acres of membrane, with pumping power sufficient to keep the hex flowing continuously through the system. The diagram (fig. 30) shows a typical arrangement of units. To add to the difficulties of diffusion plant design and operation, hex is very corrosive. It reacts fiercely with oil or grease and it combines with water to form a solid which would rapidly clog the membranes if an inward leak of moist air should occur.

The difficulties of designing, constructing and operating the first gaseous diffusion plants were at one time feared to be insuperable, but were, nevertheless, successfully overcome, as part of the wartime project in the United States, and later independently by the British, the French, the Russians and the Chinese. Details of the membranes used are still undisclosed. Diffusion plants are exceedingly expensive to build and they need vast amounts of electric power to drive the compressors, and all this power is ultimately rejected as waste heat through the cooling system.

Because of the nature of the process, a diffusion plant has to be very large to work at all. The larger the plant the cheaper the product, and at the same time, the greater the need to keep the plant continuously running at maximum output. For this reason, an assured market is needed. Since the 1960s, when the demand for weapons-grade uranium eased off, an outlet for the product was only to be found among utilities operating nuclear power plants fuelled with enriched uranium. Accordingly every effort was made by the USA to promote world-wide the construction of light water reactors committed to the continued use of enriched fuel. In the West, besides the United States, only Britain and France have (comparatively small) diffusion plants, and until recently all other countries requiring enriched fuel had, in effect, to buy from the United States, but in 1973 West Germany (through Euratom) contracted to buy from the Soviet Union. In the same year France and four other European countries agreed to form a new company, 'Eurodif' to operate a diffusion plant in competition with the United States.

The high capital and running costs of enrichment by gaseous diffusion led to widespread efforts to find less expensive and more flexible techniques in which very large scale was less important. In the early 1970s work on a high speed gas centrifuge process reached a stage where Great Britain, Holland and West Germany entered into a collaborative agreement to put such a process into action. Two companies were formed: CENTEC to design and manufacture the very large number of centrifuges that would be required, and URENCO to carry out the actual enrichment process and market the enrichment service. Production of enriched uranium started at URENCO (UK)'s first plant at Capenhurst in 1977 — the world's first commercial separation other than by diffusion. A second plant was authorised at Capenhurst in 1979 and similar plants are operating or being built in Holland. One advantage of the centrifuge process is that it is based on a large number of identical units which can be mass-produced; another is that the plant can be built to any desired scale initially, and increased in size by addition of more units according to requirements — unlike the diffusion plant which uses several different sizes of units and which needs to be built at its anticipated full size from the start. A third, and at least equally important, advantage is that for a given amount of material separated, a centrifuge plant uses only one-tenth as much electricity as a diffusion plant.

In 1972 South Africa announced that she was working on a different enrichment process which promised even more advantages. No details were released but it is believed to be based on the partial separation of light and heavy hex molecules issuing at high speed from specially designed nozzles. It too is a many-stage process.

A technique that has come on to the scene much more recently is based on the use of very finely tuned laser light, that is to say, light of a single wavelength (colour) moving in a coherent beam like a body of soldiers marching forwards in step. Laser light of a carefully selected wavelength can displace ('excite') orbital electrons in molecules containing uranium 235 more strongly than in those containing uranium 238, making them chemically more reactive than normal. If a suitable chemical reagent is also present it will react with the lighter, excited, molecules only and a clear-cut chemical separation may be achievable.

The process is arousing considerable interest because it could form the basis of a once-through separation process, operating on whatever scale of output is required. Substantial developments in laser technology and in understanding the effects of light on the chemistry of uranium will be needed before the process can be proved or applied commercially. There is some apprehension that this and other possible 'cheap' routes to the production of nearly pure uranium 235 might lead to proliferation of nuclear weapons. For this as well as for purely commercial reasons little firm information has been published about it.

The degree of enrichment required for the fuel for a particular type of reactor depends, of course, on details of the reactor design and its operational requirements. Remembering that natural uranium (as used in Magnox and CANDU reactors) contains 0.73 per cent uranium 235, the approximate upper limits in reactors already in use (see Chapter 4) are as follows: AGR, 2 per cent; steam-generating heavy water reactors, 2 per cent; American light water reactors (boiling water or pressurised water), 3 per cent; high temperature reactors of current designs, 6 per cent; fast reactors (as an alternative to plutonium/uranium 238 mixtures), 25 per cent; research reactors, up to 90 per cent or even more. Even at the low enrichment used in thermal power reactor fuels, problems associated with criticality start to arise and at high enrichments are as important with uranium as they are with plutonium.

The maximum output for any particular enrichment plant, whether based on gaseous diffusion or some other process, is measured in tonnes of 'separative work' (SW) per year. This is a concept that is not easy to express without resort to mathematics, but it takes into account the amount of enriched material produced, the extent of enrichment and the isotopic composition of the depleted material as well as of the enriched output. As a very rough rule of thumb, every 1,000 MW(e) of installed nuclear generating capacity based on low-enriched uranium requires a separative work capacity of about 100 tonnes a year to keep it in fuel. Under present (1979) plans the existing capacity of the Capenhurst diffusion plant (about 400 tonnes SW a year) is likely to meet Britain's needs for only a few more years, but it is expected that the tri-national centrifuge programme, which already produces 200 tonnes SW a year, will keep pace with the increase in nuclear generating capacity.

Turning hex into uranium dioxide fuel

The starting material for Britain's oxide fuel is slightly enriched hex from Capenhurst. This is fed as a vapour into the reconversion plant at Springfields, where it reacts in a continuous rotary kiln, first with steam to form uranyl fluoride ($UO_2 F_2$), and then with hydrogen and more steam at a higher temperature to form the black powder UO_2. The fluorine is carried away as hydrofluoric acid which is recycled to the tetrafluoride production plant described earlier. The oxide is mixed with water containing a binding material, and spray-dried to produce free-flowing granules; these are pressed into pellets and heated to drive off the binding agent, then sintered in hydrogen at $1,650°C$ so that the granules coalesce to a ceramic (pottery-like) form. Finally the pellets are ground to the precise diameter required, usually between 4 and 6 mm. The earliest designs called for solid pellets but it was found later that if they had a hole through the centre, any swelling of the fuel caused less strain on the surrounding can.

Depleted uranium

The other product of the isotope separation plants is hex depleted in uranium 235. Typically this contains only 0.2 to 0.4 per cent of uranium 235. Along with the depleted uranium recovered from the reprocessing of reactor fuel, it has the potential to become, with plutonium, one of the feed materials for the fast reactors working on the uranium/plutonium fuel cycle (see Chapter 4). Britain already holds about 20,000 tonnes of it in stock, both as hex and as oxide — an enormous reserve of energy. Apart from this potentially vital future use, depleted uranium is of very limited commercial value. The oxide has for many years been used as a pigment for ceramics, and more recently as a catalyst in the organic chemical industry, while the metal itself finds a limited outlet under the name of 'industrial uranium'. Its high density (about one and a half times that of lead) makes it valuable as a compact shielding material for medical and industrial radiation sources, or where a heavy weight is needed in a small volume, for example in inertial navigation instruments, golf-club heads or even the keels of yachts. It is also used in certain types of armour-piercing shell. The metal also shows promise as an alloying material for improving corrosion-resistance in steels and bronzes. Uranium's very slight radioactivity and its chemical toxicity constitute only minor drawbacks in these contexts: it is no more dangerous than other widely used heavy metals and its industrial toxicology is very much better understood.

6

FUEL FOR NUCLEAR REACTORS

Fuel elements — general considerations

Having obtained pure uranium metal (or oxide) of the right isotopic composition the next stage is to make it into fuel for nuclear reactors. Unlike all other kinds of fuels, which are fed in bulk to a furnace and there consumed, nuclear reactor fuel consists of a very large number of identical pieces of precision-engineered hardware, each one of which has to remain in the reactor core for several years. During this period it must retain its shape and structural integrity, while also producing tens of thousands of times as much heat as an equal amount of coal. There is no exact parallel among familiar things, but perhaps a useful analogy is the mass-produced dry battery used in an electric torch: the battery has to fit the torch for which it was designed, it must produce the specified amount of electricity to light the bulb, and it must remain physically intact until the end of its useful life. We are all familiar with the run-down and swollen battery that sticks in the torch, exuding messy and corrosive chemicals: this is the kind of fate that must on no account befall a nuclear fuel element even in the final stages of its vastly more heroic life.

Each class of nuclear reactor, indeed most of the individual designs, requires its own kind of fuel element. However, most power-reactor fuel elements in current use have a number of features in common: they each consist essentially of a uranium metal rod, or a stack of uranium oxide pellets, made to very precise dimensions and enclosed in a robust and close-fitting gas-tight metal tube or 'can'. This has as its main functions holding together and protecting the fuel itself, and preventing the fission products that are formed in it from escaping. Each element must contain just the right amount of uranium metal or oxide of the specified metallurgical quality, chemical purity and isotopic composition.

The fast neutrons produced in fission must be able to move readily, from wherever they arise in the fuel element, out into the surrounding moderator, so that they stand a good chance of being slowed down before they encounter another uranium 235 nucleus. This puts an upper limit on uranium thickness. The slowed-down neutrons must be able to get back from the moderator into the fuel to cause further fissions, without being captured by the material of the fuel can — a much more likely fate for slow neutrons than for fast ones. This limits the choice and thickness of the canning material, which in practice is virtually restricted to a few millimetres of aluminium, magnesium or zirconium or their alloys. Only if enriched uranium is used can more robust materials, including stainless steel, be employed.

The heat produced by fission must be able to get away through the fuel and the can walls into the coolant-fluid fast enough to produce power at the required rate, and to prevent the fuel from overheating. This calls for good thermal conductivity in the fuel (as is the case with metal) or a more slender design of fuel element as is needed with oxide fuel of lower conductivity. Thermal contact between the fuel and the can wall must be good, which usually means that the can must be pressed into very close contact with the fuel, and any gaps filled with a good conductor of heat, such as helium gas. To increase the flow of heat the fuel elements have finned or ribbed outer surfaces, which not only extend the area of contact with the coolant fluid but also create turbulence in it, which aids heat transfer.

Unlike neutrons and heat, the fission products which gradually accumulate in the fuel must on no account be allowed to escape from it. It is a primary function of the can to prevent this throughout the lifetime of the fuel element. There must be no flaws in the can itself or its means of closure, nor must any be allowed to develop in service. This can only be achieved by careful design, tried out thoroughly in test rigs and research reactors, with the most rigorous inspections at all stages of manufacture. The fuel itself must play its part by helping to retain the newly-produced fission-product atoms within its own crystalline structure. This means that every time a fission takes place room must be found in the fuel for two atoms where previously there was only one, so there must be some internal 'give' in its structure. At the same time any tendency on the part of these new atoms to tunnel their way out of the fuel or (in the case of gaseous fission products) to accumulate as growing bubbles of gas, must be guarded against. In the case of uranium metal, which is more susceptible to this trouble than the oxide, the problem is best tackled by judicious alloying and heat treatment during manufacture, so as to produce the right metallurgical grain structure. Repeated heating and cooling of the fuel in use, due to shutting down and restarting the reactor, adds to the difficulties and is one of the reasons why the Magnox power stations work better if kept on steady 'base-load' operation than if they attempt continually to satisfy changing power demands.

The fuel and canning materials must be compatible with one another and there must be no tendency for them to run together or form alloys at the temperature of operation, so that weak points are formed leading to possible flaws. Similarly, the canning material must be compatible with the coolant fluid and with the moderator, otherwise corrosion may occur. The position can become quite complex: for example, in gas-cooled reactors (Magnox and AGR) the carbon dioxide coolant under the combined effects of heat and radiation can attack the graphite moderator and deposit carbon on the boiler-tube surfaces. This can erode the graphite and at the same time interfere with the transfer of heat. An extensive programme of research was needed to find the remedy, which lay in the addition of carefully controlled amounts of other selected gases to the carbon dioxide coolant to inhibit the reaction.

The fuel element must be mechanically strong, and remain so. It has to withstand initial handling and insertion into the reactor core, several years of intense neutron bombardment at a high temperature and often under heavy mechanical stress, including contact with a strong and turbulent flow of hot gas or liquid. The bottom fuel element in a vertical channel may have to carry the weight of all those above it without bending: in some Magnox reactors this amounts to 10 kilograms per square centimetre. This mechanical strength is usually provided by the can rather than the fuel itself, and additional bracing pieces or sleeves may be fitted to the outsides of the can to stop them bowing, and these may also serve to keep the fuel elements aligned centrally in the channels.

Later, the fuel may have to suffer further mechanical stresses when it is moved from one place to another within the reactor while it is still hot, and on removal at the end of its life from the reactor to the cold water of the storage pond, where it may remain for several months — years, even — prior to removal and transport to the reprocessing plant.

In some reactors, for example Magnox, the fuel elements will be in the form of rods which are handled singly, while in others, for example AGR, it will consist of 'pins' of much smaller diameter, grouped in clusters for structural strength and convenience in handling. These clusters may be further assembled end to end in 'stringers', one in each vertical channel. In either case, provision must be made in the construction of the fuel elements for 'blind' handling with mechanical grabs.

Finally, but of vital importance to the practicality or otherwise of nuclear power, the fuel elements must be capable of being manufactured to the strictest standards in very large numbers at a commercially acceptable cost.

Fuel element production – Magnox

We will now look in some detail at the production of fuel for use in the reactors of a typical Magnox power station – the Central Electricity Generating Board's 550 MW(e) power station at Sizewell 'A' in Suffolk. Each of the two reactors contains some 35,000 individual fuel elements all made to the same design specifically for this station. Fuel elements for other Magnox stations differ in detail, mainly in regard to external fittings and overall length: the principles are the same for all.

Each element consists basically of a 12 kg uranium metal rod about 1 metre long by 3 centimetres in diameter, enclosed in a finned Magnox can. Each rod gives out the same amount of heat as about thirty single-bar electric fires. Most of the rods remain in the reactor for upwards of three years, during which time each one can be expected to produce enough heat to generate, on average, about a million kilowatt hours of electricity. This is equivalent to the burning of about 150 tons of coal – in Britain roughly one person's lifetime energy needs, domestic and industrial.

The first stage in manufacture is to melt a billet of uranium and add the alloying materials needed to give the right metallurgical qualities. Heating is by induction, that is to say by surrounding the graphite crucible with water-cooled coils through which a heavy high-frequency electric current is passed; the induced 'eddy-currents' melt the uranium which is then poured into a set of cylindrical steel moulds and allowed to solidify. Because uranium metal is oxidised very quickly at high temperatures, the whole melting and pouring operation has to be carried out in a high vacuum.

The cast rods are not yet ready for final machining because during cooling the uranium atoms have had time to arrange themselves in well-defined crystals, the largest at the centre of the rod where cooling was slowest. If the rods were to be used in this state the larger crystals would grow under the influence of neutron bombardment and the rod would gradually lose its shape. In addition, the fission product atoms would tend to collect together in the boundary areas between the large crystals and cause swelling and blistering. By strongly re-heating the rods and rapidly 'quenching' them in cold water the structure of the metal is changed to one of smaller, uniform crystals pointing in all directions and much less subject to the troubles just described. To relieve any internal stresses the rods are annealed by heating them yet again, slowly this time and to only about 550°C, and letting them cool naturally.

They are now ready for machining to their exact size, after which a number of shallow grooves are cut around each rod at intervals along its length. These help to hold the rod and can firmly together despite their different rates of expansion under heat, which would otherwise lead to progressive stretching and possible rupture of the can by 'thermal ratcheting'. After careful inspection the rods, numbered serially for reference, are now ready for putting into their cans (see fig. 31).

The cans are made from a specially developed alloy of magnesium called 'Magnox' ('magnesium, no oxidation'). This has the required

Fig. 31   *Part of the fuel canning line at the Springfields works. Each Magnox-clad uranium rod can supply energy equivalent to 150 tons of coal. (Courtesy of BNFL.)*

strength and resistance to corrosion by hot carbon dioxide gas under powerful neutron bombardment, and retains its good 'transparency' to thermal neutrons. Purity and metallurgical quality are almost as important in the canning material as in the fuel itself. The cans themselves are quite elaborate engineered products and they have to be subject to stage-by-stage inspections and controls as rigorous as those applied to the fuel rods themselves.

The rods are inserted in the cans under conditions of almost surgical cleanliness, well away from areas in which uranium is being handled. This is because even tiny traces of uranium on the can surfaces could give rise to fission products in the coolant gas, and these would trigger off the very sensitive alarm system that monitors the reactor for leaks in the fuel elements. Residual air is pumped out of the can and replaced by helium gas (which is a good conductor of heat and easy to detect if it leaks); the end-caps are welded on automatically under computer control, using an arc welding technique specially developed for the purpose. The can is then forced into very close contact with the uranium rod by all-round hydraulic pressure, to ensure good heat transfer and to prevent thermal ratcheting as described above.

Each fuel element is submitted at this stage to very critical scrutiny by expert inspectors, by X-rays and by a 'mass-spectrometry' leak test based on the detection of escaping helium. After careful cleaning each element is fitted with all its appropriate accessories including end fittings, centering devices, braces to strengthen and prevent bowing in the reactor, gas-flow 'splitters' and anything else that the particular design may call for, such as support rods, graphite sleeves, etc. After careful final inspection the completed fuel elements are individually numbered, wrapped in plastic sleeves and packed in stout boxes for despatch to the reactor site. Transport of this class of reactor fuel presents no special problems or hazards: in the UK it is sent by the most convenient road or rail route. On receipt at the power station the fuel elements are again thoroughly inspected before acceptance, and yet again before use: nothing is left to chance or taken on trust. Such extreme care pays dividends in terms of product reliability and operational safety: out of over three million Magnox fuel elements supplied from Springfields during twenty-five years, only a tiny number have developed defects in service which required their removal from the reactor before the scheduled date. Except for some experimental fuel elements which developed leaks when undergoing tests, there have been no fuel-element failures which necessitated the unscheduled shut-down of a UK power station reactor.

Oxide fuels

The oxide fuels for AGRs and water reactors are subject to many of the same design and production considerations as Magnox fuel, so only the major differences will be described here. Britain's AGRs are designed to generate steam hot enough to drive the standard 660 MW turbo-alternator sets of our most up-to-date coal-fired power stations. This means that the fuel elements must permit gas outlet temperatures of about 650°C, which is well above the softening points both of uranium metal and of Magnox cladding material (gas outlet temperature in the latest Magnox reactors is only about 410°C). However, uranium oxide, UO_2, can be manufactured in a very stable ceramic form which will easily withstand this temperature and the other conditions in which it must operate. Oxide fuel, though, has two major drawbacks: the oxide does not conduct heat as well as the metal, and the fuel is less concentrated. The first of these is easy to overcome by reducing the diameter of the fuel elements from about 30 mm to about 6 mm so that the heat has less far to travel to get from the inside of the fuel to its surface, while at the same time, the surface area available for contact with the coolant gas is increased in relation to the mass of fuel. The two-to-one dilution (in terms of numbers of atoms) of uranium with oxygen reduces the chance of a neutron striking a uranium 235 nucleus, so special attention has to be paid to keeping other neutron losses to the minimum, especially in the canning material. At one time, it was hoped that beryllium, which has very low neutron absorption and resists high temperatures, would be found suitable for this, but it proved too difficult to use and too costly. At the expense of neutron economy, stainless steel with its excellent resistance to heat and chemical attack was chosen instead. To make up for the added neutron losses, the uranium in the fuel was enriched to a U-235 content of 1½ to 2½ per cent, i.e. about two to three times its natural abundance. Although this puts up the cost of the fuel substantially, it gives better 'burn-up', that is to say it enables a lot more energy to be got from a given amount of starting material (natural uranium), so it has advantages in its own right in addition to compensating for poor neutron economy.

Britain's AGR fuel elements, unlike Magnox, are of the same design for all the commercial reactors. In making AGR fuel the oxide pellets are automatically inserted under clean conditions into thin-walled stainless steel tubes about 1 metre long and with a slightly corrugated surface to aid heat transfer. The loaded tubes or 'pins' are filled with helium gas and the end caps welded on; each pin is leak-tested and subjected to hydrostatic pressure to ensure good internal heat transfer. After cleaning and inspection the pins are assembled into clusters of 36 and fitted with appropriate end-plates, spacers and other external fittings. Finally, a double-walled graphite sleeve about 24 cm across is fitted over the whole length of the cluster to form the complete fuel element (see fig. 32). Each element contains about 43 kilograms of slightly enriched uranium oxide and gives a heat output during its three to four year lifetime in the reactor of about 20 million kilowatt-hours of electricity. So this single piece of hardware, about a metre long and 24 centimetres across, is equivalent in its usable energy content to about 3,000 tonnes of coal. Each of Britain's ten AGRs is designed to take a charge of just over 1,000 such fuel clusters, joined end to end in 'stringers' of six clusters, one to each vertical fuel channel.

The fuel for water reactors − PWRs, BWRs, CANDU and

Fig. 32 *Fuel elements for the Advanced Gas-Cooled Reactor. The stainless steel tubes contain slightly enriched uranium oxide pellets and are surrounded by a graphite sleeve. Each bundle can supply energy equivalent to burning 3,000 tons of coal. (Courtesy of BNFL.)*

Britain's SGHWR — also takes the form of tubes filled with suitably enriched uranium oxide pellets. However, because water is present in the reactor core as coolant or moderator (or both) the outlet temperature from the core is limited to only about 280°C. Stainless steel is not necessary and the fuel is canned in a zirconium-aluminium alloy called 'Zircaloy'. This withstands the conditions in the reactor and does not absorb nearly as many neutrons as stainless steel.

7

THE FUEL IN
THE REACTOR AND AFTER

Fuel performance

Unlike a coal- or oil-fired power station furnace, a nuclear reactor is filled to capacity with fuel before it ever comes into operation, and this initial fuel charge is expected to keep it going for a period of several years before replacement starts. The fuel itself must therefore be dependable upon to behave predictably throughout its life in the reactor, and future supplies of replenishment fuel of the same specification must continue to be available for the whole lifetime of the reactor, which may be as long as thirty years.

The first fuel charge

Loading the first fuel charge (see fig. 33) forms part of the commissioning procedure for the reactor and follows a carefully worked out schedule that takes into account the underlying physics of the nuclear chain reaction and the detailed design of the particular reactor. Briefly, fuel loading starts from the centre of the core and proceeds outwards, with all control and shut-down rods fully inserted. Artificial sources of neutrons (see Chapter 3) are sited at strategic points and the way in which the neutron populations build up or die away as loading proceeds is closely watched and compared with the design figures. This becomes especially important at the stage when criticality is approached, that is to say, when the chain reaction is about to become self-sustaining.

When loading is completed satisfactorily the reactor is gradually run up to power by withdrawing first the shut-down rods and then the control rods, while continuing to keep a close check on neutron populations and their rates of growth, also on temperatures in fuel, moderator, coolant, and a great many other important variables. Commissioning may take a year or more even when everything goes according to the designers' expectations. During this period all the components and equipment of the reactor itself, and

Fig. 33   *Preparing to load a consignment of Magnox fuel into one of the Calder Hall reactors. The anti-ratcheting grooves into which the Magnox cans have been pressurised show clearly on the left-hand elements. (See Chapter 6. p. 100).*

of the associated steam and electricity generating systems are thoroughly tested, not only individually, but as parts of a very complex integrated system. Thereafter, the whole system must work smoothly as a single unit for its full lifetime. Unlike a coal or oil furnace which can be let out, cooled down, and made freely accessible inside, the interior of a nuclear reactor that has once been in operation becomes too radioactive ever again to be entered except on a very restricted basis.

As the fuel gradually yields up its energy in the reactor, the slow but important changes that take place in it lead to gradual long-term changes in reactor behaviour. For example, fissionable uranium atoms become less plentiful as they take their turn in the chain reaction, while fission-product atoms and the products of their radioactive decay increase in number and variety and in their capacity to capture neutrons. At the same time fissionable plutonium builds up and produces its own fission products, while in the case of oxide fuels especially, there is also a build-up of 'actinide' elements heavier even than plutonium. All these changes, and many others, have to be foreseen in detail by the reactor designers and taken into account by the operators.

After the reactor has been running for a period of some months or years, some of the fuel will no longer be contributing its share of power but will be starting to act more as a trap for neutrons, so the complex business of refuelling will have to start. Depending on the design of the reactor and its fuel this may be undertaken either as a continuing operation with the reactor on load, or while the reactor is shut down for maintenance every six months or so. Britain's commercial Magnox and AGR stations are designed for on-load refuelling (see Chapter 4). Some of the fuel that has been most fully burned up (usually in the central region of the core) is the first to be removed from the reactor and sent for reprocessing. It is replaced by new fuel, or more usually by fuel already in the reactor but located where the chain reaction has been less intense, and fresh fuel is put into the vacant places. By 'shuffling' the fuel in this way the rate at which fission takes place is kept as uniform as possible throughout the volume of the reactor core, and with it the rate of heat output.

Storage and transport of spent fuel

After a fuel element has reached its maximum useful life in the reactor, with or without reshuffling, it is removed and parked in a shielded and cooled vault. From here, batches of fuel are periodically removed, still under heavy shielding, to the power station's 'cooling pond' where, still in its can but stripped of other extraneous hardware, the fuel is stored under the safe and transparent shielding provided by several metres of water. Here it remains for a period that ranges from about six months in the case of Magnox fuel, to several years for the more robustly clad oxide fuels, while the fission-product radioactivity dies down to a level at which the fuel can be taken away to the reprocessing plant without the additional inconvenience of having to cope with excessive heat from its radioactive decay. The fuel elements are loaded, still underwater, into massive lead and steel transport containers or 'flasks' (see fig. 34). These are built not only to keep the level of external radiation down to a prescribed safe figure, but also to withstand the severest accident that they are likely to encounter *en route*, without losing their shielding efficiency or allowing radioactive materials to escape. To check the design stringent tests are carried out under internationally agreed standards, including successively a free drop of

Fig. 34   *Flasks containing spent nuclear fuel being transported from a power station to the Windscale Reprocessing Plant in Cumbria. (Courtesy of BNFL.)*

9 metres onto solid rock or concrete and 30 minutes in blazing oil, also a water immersion test. The flasks are fitted with external cooling fins to help disperse the radioactive decay heat that is still given off. In the UK the flasks are moved, where possible by rail, from the power stations to the reprocessing works at Windscale in Cumbria. On arrival they are off-loaded into the central cooling pond and the fuel elements removed for a further period of underwater storage. This usually brings the total cooling-off time to between one and two years for Magnox, longer for oxide fuels. By this time the activity due to the shorter-lived fission products (especially the important iodine isotopes) has virtually disappeared. Very substantial radioactivity, however, still remains (together with quite a big output of heat), but this is much slower to die away and little is to be gained by putting off reprocessing for any longer. Magnox fuel will not in any case stand up to prolonged underwater storage.

Reprocessing

In an economy where uranium is plentiful in relation to the forecasts of energy requirements, such as Canada and the United States, a strong case can be made out for not reprocessing the spent fuel at all. It may be regarded instead as waste to be disposed of with no prospect of retrieval or it may be kept in safe storage with the option of deciding later whether to reprocess it. The Canadian CANDU reactor system was conceived and continues to be operated on this basis. The United States are also well placed for uranium supplies and have no urgent need for reprocessing, and where this has been attempted it has met with little commercial or technical success. In the late 1970s it had become US Government policy to discourage reprocessing of water-reactor fuels for the time being and to slow down on some aspects of the development of fast reactors.

However, a no-reprocessing system is potentially very vulnerable to adverse changes in uranium supply and price, whereas systems which incorporate fuel reprocessing, with recycling of uranium and plutonium through fast reactors, are much less affected and may indeed eventually become almost completely insulated from world uranium price changes. Further, to store very large numbers of spent fuel elements for an indefinitely long time poses a great problem, not only in physically isolating them and the radioactivity that they contain so that they can do no harm through leakage or accident: they also form a store of plutonium which year by year grows larger in amount and less well-protected by the deterrent effect of radiation (from the accompanying fission products in the fuel) and therefore more of a temptation to would-be bomb makers (see Chapter 12).

In Britain reprocessing for military needs preceded the advent of nuclear power by some five years, and there has never been any serious question of not reprocessing the fuel from the commercial Magnox stations: the plant was in successful operation already. Furthermore, quite independently of plutonium recovery for military use, Britain, the United States and the Soviet Union have from the earliest days seen plutonium-burning fast reactors as the only truly economic way in the long term of exploiting nuclear energy to the full. A fast reactor system requires a substantial initial stock of plutonium before it can start, also a supply of depleted uranium to keep it going. But reprocessing is a difficult and demanding technology which not many countries are able, or indeed willing, to undertake.

France, which after starting with gas-graphite reactors akin to Magnox later switched to LWRs, built a large reprocessing plant in Normandy. This can also undertake work for overseas customers including Japan – competing in this with the British THORP plant (Thermal Oxide Reprocessing Plant) under construction at Windscale (see p. 115). Japan, as well as placing big reprocessing orders overseas, is building a pilot reprocessing plant of her own on which to base the capacity for meeting the needs of her very large projected nuclear power programme, which is expected eventually to include several fast reactors.

The main practical factor that makes reprocessing difficult is the intense and penetrating radiation given out by the fission products, even after the fuel has spent many months in the cooling ponds. Shielding and other measures have to be adequate not only for normal operating and maintenance, but also in the event of accident. Additionally, the presence of plutonium which is both fissile and radioactive needs its own special measures – which in recent years have come to include a new emphasis on physical security.

The treatment of spent nuclear fuel at the Windscale reprocessing plant has much in common – and much that contrasts – with the chemical purification of uranium oxide at Springfields. The Windscale process, however, is very much more complex chemically – there is more than one end product to be considered – and it is made more difficult by major considerations of radiological safety that do not apply at Springfields.

All plant that contains unseparated fuel or significant amounts of fission products must be effectively sealed to stop radioactive materials escaping and, at the same time, it must be surrounded by up to 2 metres of concrete biological shielding, behind which all operations must be remotely controlled. Once it is operational, access can only be gained in a carefully planned procedure, and routine maintenance in the ordinary sense is not possible for the rest of the plant's working life. The need for access must therefore be eliminated as far as possible. Virtually the whole plant is constructed from heavy-gauge stainless steel, all joints are welded to the highest standards and radiographed both as an inspection check and as a permanent record; gravity feed, air pressure or vacuum are used instead of pumps, and metering vessels that automatically fill and discharge themselves replace normal flow-metering. All plant controls and instruments for measuring temperature, pressure, liquid level, acidity, etc., transmit their signals (in many cases by pneumatic or hydraulic means) to the control areas on the safe side of the shielding. Facilities are installed for sampling the process material at key points so that specimens for analysis can be delivered automatically or on demand. Detailed records are kept of day-to-day operations, and these have proved their value in the control or prevention of malfunctions, in the investigations of plant incidents and in the development of improved techniques and new designs.

The main operations, whether for metal or oxide fuel, comprise preparation, dissolving, separation of the solution into the three main streams (uranium, plutonium and fission products) and the further treatment of each of these materials in preparation for their re-use, stockpiling or disposal. Let us deal with Magnox fuel first.

The metal rods are stripped of their cans and fed one at a time into a dissolving-vessel of hot nitric acid with other chemicals to aid dissolving and separation (see fig. 35). Various gases are given off, including oxides of nitrogen which are recovered and re-cycled as nitric acid, and some radioactive krypton 85 from the fuel, which is for the time being authorised to be released to the air. The filtered liquid from the dissolver flows by gravity to conditioning tanks and thence to the primary separation plant which uses horizontal mixer-settlers like those at Springfields but smaller and specially designed for maintenance-free operation, and the same solvent, tributyl phosphate mixed with kerosene. In the first

Fig. 35   *The fuel reprocessing plant at Windscale. View on top of the cells in which highly radioactive spent Magnox fuel is chemically separated. (Courtesy of BNFL.)*

part of the plant the uranium and plutonium nitrates go together into the solvent, leaving the bulk of the radioactive fission products in the acid. In the next part the uranium and plutonium are washed back from the solvent into water. Chemicals are again added so that on entering a third set of mixer-settlers the uranium goes into the solvent, while the plutonium remains in the water.

The final purification of the uranium and plutonium solutions, from each other and from residual fission products, runs on much the same lines as the primary separation process, but using different additives and different operating conditions. The solvent is recovered, cleaned up and recycled with very little loss.

Plutonium

Plutonium brings in its train added complications associated with criticality (see Chapter 1 p. 20) and radio-toxicity. Any accumulation of plutonium above a certain mass may lead to the unintentional start of a fission chain reaction with a sudden and potentially dangerous release of energy and radiation. This must on no account be allowed to happen. The design of all plant and equipment in which plutonium is handled or stored must be such that in no credible circumstances can such an accumulation arise, whether accidentally or by plant malfunction or breakdown or by human error. The size and shape of plant and the presence nearby of neutron-moderating materials are all subject to design restrictions, backed up by operational discipline and strict accounting. The hydrogen and carbon atoms of a glove-box operator's hands can slow down neutrons almost as effectively as the graphite moderator in a reactor: if enough plutonium is present, the same sort of chain reaction can ensue.

Plutonium is also dangerous if absorbed into the body, so it must be processed and stored in such a way that it cannot be ingested, inhaled or brought into contact with skin. Before its separation from the other constituents of the spent fuel, the toxicity of plutonium is overshadowed by the much greater activity of the fission products, and the measures taken to safeguard these take care of the plutonium also. However, when it has been separated for purification and reprocessing, plutonium again calls for special care. It no longer needs heavy shielding because the only significant radiation it emits is alpha-particles which have very low penetrating power: a sheet of plastic or a pair of rubber gloves is enough. But it still needs to be effectively contained and isolated from the environment in which people work, and in particular it must not get into the air that they breathe. For this reason work with plutonium is done in sealed compartments or glove-boxes, as described earlier. In a production plant, a number of these boxes may be connected together by sealed ports through which the active materials can be passed from stage to stage of the process. This, together with considerations of criticality, limits the scale of operations on any part of the production line. It is no great disadvantage because only a few tonnes a year are handled (see p. 131).

The plutonium nitrate solution that arises from the main separation plant is purified by further solvent separation processes and precipitated as an ammonia compound which is heated to give the pure oxide PuO_2. Ground and mixed with about four times its weight of pure uranium dioxide (natural or depleted), this forms the material from which fast reactor fuel is made (see p. 73).

A promising alternative way of making the mixed oxide fuel is by the 'gel precipitation' route: a solution containing the right portions of uranium and plutonium nitrates is introduced into an alkaline solution containing a colloidal (jelly-forming) material, so that little globules of mixed oxide jelly are formed. These can be controlled very closely in size (about 0.1 to 1.0 mm in diameter), texture and composition and after drying are very well suited for making into fast reactor fuel. The process has the advantage that it does not at any stage give rise to plutonium-containing dust.

Uranium recovery

The second main product from the primary separation plant is depleted uranium. This is freed from residual fission products and plutonium by further separation processes and converted to oxide for later use in fast reactors or, because it still contains useful amounts of U-235, to hexafluoride for the isotope enrichment plant.

Wastes

The third main product from the separation plant is the radioactive fission-product waste. Besides fission products with half-lives ranging usually between a few weeks and thirty years, this also contains a small amount of plutonium, but this is almost completely recovered in a further separation process – it is too valuable to discard. Some uranium goes with the fission products, but this is unimportant because it adds nothing to their hazard and is not valuable enough to warrant the added cost and handling effort of its more complete recovery. Waste treatment is described in Chapter 8.

Reprocessing oxide fuels

The process just outlined has been used satisfactorily at Windscale for three decades for reprocessing unenriched uranium metal fuel. It has handled not only the military production of the Windscale reactors, but also all the irradiated fuel from Calder Hall and Chapelcross as well as from Britain's nine commercial Magnox power stations and from two similar stations built by Britain in Italy (Latina) and Japan (Tokai Mura) in the early 1960s. But the Advanced Gas-Cooled Reactors (AGRs) of Britain's second and third nuclear power programmes are based on slightly enriched uranium oxide fuels (as are most of the power reactors — BWRs and PWRs, — currently operating overseas). Basically there is little difference in the reprocessing technology because nitric acid dissolves uranium metal and oxide almost equally readily and thereafter the separation processes are the same, although experience has shown that oxide fuels are, in fact, more difficult to reprocess than metal fuels.

The major difficulties arise from the fuel element construction and from the much higher burn-up achieved by oxide fuels. A preparatory head-end section strips the fuel elements of extraneous hardware and chops the pins into short lengths. A dissolver brings the uranium and its irradiation products into solution in nitric acid while leaving the stainless steel or zirconium alloy canning materials undissolved.

The high burn-up of oxide fuels leads to a higher proportion of the heavier isotopes of plutonium being produced, and these decay to give very long-lived alpha-emitting isotopes of the actinide elements curium and americium, which call for special consideration in long-term waste management. The high burn-up of the fuel also affects the behaviour of the fission products. Some of these, notably ruthenium, tend to migrate within the crystal structure of the uranium oxide while it is in the reactor and to form lumps which are hard to dissolve. In 1974 an unsuspected accumulation of this material in a part of the Windscale oxide reprocessing plant led to a chemical reaction and a blow-back of air contaminated with ruthenium into the shielded working area; the plant had to be closed for repairs, and altered to ensure that the same thing could not happen again. This occurrence had nothing to do with criticality. The section of the plant dealing with Magnox fuel was not affected and continues to operate.

However, oxide fuel reprocessing had to stop and AGR fuel, together with fuel from overseas, continues to be stored in the cooling ponds at Windscale. The stainless steel cladding of AGR fuel and the zirconium alloy of water reactor fuels can stand underwater storage for up to ten or twenty years respectively. To handle fuel from Britain's AGR stations (and from any water reactors that we may build), and from overseas customers, a new plant is required. Accordingly, plans were drawn up by BNFL for a complete new plant (THORP) to be constructed alongside the existing plant at Windscale, to come into operation in the 1980s. Some of the money to build THORP would be found by Japan as a major prospective customer, the rest being raised by BNFL.

The Secretary of State for the Environment instituted a full public inquiry under Mr Justice Parker who in 1978 recommended acceptance of the proposal subject to certain conditions, and Parliament confirmed this. It was made clear at the Inquiry that the regulating authorities would not necessarily permit any increase in radioactive discharges at Windscale when the new plant opened: BNFL was able to give satisfactory assurances that increased discharges could be avoided. The plant is due to be completed in the 1980s and, with a nominal throughput capacity of up to 1,200 tonnes a year, will be able to handle all the foreseeable needs of the United Kingdom together with substantial overseas work.

8
RADIOACTIVE WASTES

Waste problems

All industries, and most other human activities, give rise to wastes, some of which can be dangerous to people, damaging to the environment or expensive to get rid of satisfactorily. For example, the chemical and mining industries give rise to effluents and 'tailings' which may be unsightly or poisonous, while farming, the food processing industries and man himself produce wastes that may be unpleasant or biologically harmful unless properly treated; in a number of cases the problems are still far from being solved. For example toxic metal residues (lead, cadmium, zinc, mercury), slurry from intensive animal farming, sulphur dioxide from coal-fired power stations, and of course household refuse in large cities, all need further study.

The nuclear industry is in a paradoxical situation regarding waste: in spite of its importance to the nation's well-being, it handles only a few thousand tons a year of nuclear fuel, nearly all of this in three factories and about a dozen power stations. Most of it is recycled or stock-piled, and only a tiny fraction of it becomes 'waste' in the true sense. However, this tiny fraction calls for special consideration because it is potentially harmful and arouses strong emotional reactions. Radiation and its dangers were well understood even before Britain's nuclear power industry was born: its founders knew quite well that they would have a very special waste problem to tackle, and they already had a useful basis of knowledge from which to work, together with a 'feel' for the subject and respect for the problems that it would bring.

Not all of the wastes arising in the nuclear industry are radio-active in the sense of being more radioactive than the water, air and soil that we are accustomed to living with. Waste water, scrap machinery, ordinary laboratory wastes, paper, textiles, and some chemical wastes — provided that monitoring has shown them to be 'inactive' — may be treated as ordinary domestic or 'trade' waste and disposed of through the usual channels approved by the local authority: they present no special nuclear problems.

Radioactive wastes, that is to say unwanted radioactive materials, arise at several places in the nuclear fuel cycle, but before looking at these in detail let us see where the radioactivity itself comes from. It has three main sources:

1. Radioactive impurities present in the ore from which the uranium fuel is derived and from which the uranium has to be freed: these arise at the Springfields uranium purification plant (and, of course, at the place of origin in mine and mill).

2. Newly-created radioactivity produced when neutrons hit materials and objects within or around the reactor core, including the fuel and its cladding. Plutonium is produced in this way, but it is in no sense a 'waste'.

3. By far the most significant, the fission of uranium itself (and of plutonium) resulting in the build-up of a large range of fission-products of intermediate atomic weights, all of them radioactive and some decaying in stages through a series of radioactive daughter-products.

Wherever they originate, the radioactive wastes can all be classified according to the level of their radioactivity as high-level, medium-level and low-level wastes, and they may be in solid, liquid or gaseous form. Some are short-lived, their radioactivity dying away over a period of days, weeks or months, while others remain significantly radioactive for years or even centuries. A very few remain active for tens of thousands of years. Those that are the slowest to decay are for that very reason the least active though they are not necessarily the least significant.

In principle, radioactive waste can be dealt with either by dispersing it to the environment in such a way, or at such great dilution, that it presents no appreciable risk to people, or by containing it in a safe place for long enough for the radioactivity to have fallen to a level at which there is a similar absence of risk.

Control over the discharges of radioactive wastes in the UK is governed by the Radioactive Substances Act 1960 which prohibits the disposal of radioactive wastes on or from all premises — power stations, for example — unless authorised by the appropriate

Ministers. (Although Crown premises are technically exempt they nevertheless have to conform to the same standards.) The objectives of controlling the disposal of waste into the environment are laid down as follows:

1. Irrespective of cost, to ensure that no member of the public receives more than the dose limits recommended by the ICRP* and to ensure that the population of the country does not receive an average dose of more than 1 rem per person in thirty years (see Chapter 7 p. 131).

2. Having regard to cost, convenience and the national interest, to reduce doses as far below these levels as is reasonably practicable.

Authorisations for disposal of radioactive waste into the environment are granted only after the authorising Departments are satisfied that the operator has a justified need to dispose of the waste, with the overriding requirement that the disposal will not lead to any of the ICRP-recommended dose limits for exposure of members of the public being exceeded.

Low-level wastes

Low-level liquid wastes — large volumes of water containing very small amounts of dissolved radioactive substances — arise at the chemical plants where nuclear fuels are manufactured or reprocessed, at nuclear research laboratories, and at places where spent nuclear fuel is stored underwater pending reprocessing, that is to say at Windscale and at the nuclear power stations. The most important source of these wastes in the UK is the reprocessing plant at Windscale where large quantities of water are used at various stages in the chemical separation plant and in the purification lines for plutonium and recovered uranium.

After being impounded for some months to allow short-lived radioactivity to die away, the water is treated to remove as much of the remaining dissolved radioactive materials as is practicable, then discharged to the sea-bed on a falling tide through a pipeline stretching two miles out to sea. The quantities of each separate radioactive substance that may be discharged over a fixed period are closely specified in the authorisations issued by the Department of the Environment, and careful independent checks are made to

International Commission on Radiological Protection (see below).

ensure that the conditions are complied with, also to find out whether there is any unexpected build-up of radioactivity in the sea, or on the neighbouring beaches or in marine organisms. The discharge limits ensure that any radioactive material finding its way through a 'food-chain' (for example, via plankton, shellfish and flat-fish) to man will be in quantities well below the recommended limits of human intake set by the International Commission on Radiological Protection — limits which are accepted by all countries (see Chapter 8). These limits are interpreted in the UK by the National Radiological Protection Board in terms of the amounts of each substance that may be discharged. Actual discharges are only authorised at very much lower levels.

Slightly radioactive water discharged periodically from the fuel-element cooling ponds at Windscale and at the nuclear power stations is controlled on similar lines, as are the laboratory effluents from establishments such as the Atomic Energy Research Establishment at Harwell — which discharges its low-level effluent into the Thames below Oxford. (It is interesting to note that the water discharged by Harwell is chemically purer than the river water itself, and the water below the discharge point is less radioactive than the mineral waters at Bath — which people have been taking for many centuries to improve their health.)

Besides the low-level liquid wastes discharged to rivers or the sea, there are also some long-lived gaseous wastes, the predominant one of which is the fission-product gas krypton 85. At present they are all discharged, under authorisations, to the atmosphere, but if and when nuclear power programmes in Britain and the world are greatly increased, it will probably be considered advisable to restrict their discharges. Methods are being developed, for example, to implant the krypton electrically, atom by atom, in the crystal lattice of a metal where it will be held safely until its radioactivity (which decays with a half-life of about ten years) is low enough for the implanted metal blocks to be treated as medium-level solid waste (see below).

Low-level solid wastes, scrap hardware, etc., provided that they do not contain more than a very little plutonium, are buried at shallow depths on enclosed sites, or they are embedded in drums of cement and disposed of in a carefully chosen area of the deep ocean under international agreement and supervision. Radioactive decay will outstrip their possible re-entry into man's environment or food-chain. Excavated caverns, salt-mines, etc., are used for controlled storage in some countries.

Medium-level wastes

These include laboratory and plant scrap, the discarded cladding and components of reactor fuel elements from the input end of the reprocessing plant, and some of the chemicals arising in the separation and purification processes, also low-level scrap containing plutonium. Generally speaking, they are kept under surveillance in appropriate stores on the sites at which they arise, where their radioactivity decays, eventually reaching low levels at which disposal is appropriate. Fuel cans and other extraneous hardware removed before the fuel is reprocessed are stored underwater in concrete silos at Windscale. Unlike the high-level wastes described below the heat that they produce does not lead to any significant problems.

High-level liquid wastes

This is the major waste arising from the reprocessing plant itself, carrying practically all the fission products separated from the spent fuel, and it is in most respects the most important waste of the nuclear industry. It is a 'waste' in the true sense of being what is left of the uranium or plutonium atoms after they have split in two and yielded up their energy, and are good for nothing more. The amount produced is directly related to the amount of the fissile material used up and (taking account of the varying efficiency of generation) to the amount of nuclear electric power produced. The composition of the waste is affected only slightly by its origin, whether from slow-neutron fission of uranium 235 in Magnox, AGR or PWR, or from the fast-neutron fission of plutonium in fast reactors. Nor is it much affected by the degree of burn-up of the fuel in the reactor (i.e. the energy that it has yielded per tonne of fuel used). Many of the fission-product elements have half-lives so short that they are no longer active when they reach the separation plant. However, about nine chemical elements having half-lives between eight days and thirty years are still present in substantial amounts in the wastes arising from the plant. The same stream also contains small amounts of the heavy, alpha-emitting actinide elements (neptunium, americium and curium) with traces of residual plutonium, and uranium. Some of these have exceedingly long half-lives, reaching thousands or tens of thousands of years; eventually they become identical with naturally occurring radioactive elements and follow the same patterns of radioactive decay.

So much energy is released in the radioactive decay of the fission products mixture that the temperature of the waste liquid rises substantially, and it is this heat output almost as much as the radiations themselves that present problems in handling and storing the liquid. At Windscale the present practice is to evaporate the liquid down to about one-fifth of its volume (using its own self-heating properties) then to pump it into storage tanks of special design. These tanks are built from stainless steel, they are double-walled, and each is surrounded by a concrete radiation shield with a stainless-steel-lined catchment basin. The contents of the tanks — an acid solution containing undissolved matter, in suspension — is continually stirred to prevent it settling out, while it is prevented from overheating by a number of independently operated cooling-water coils. Elaborate leak detection systems operate in the space between the double walls and in the catchment basins of each tank, and provision is made to pump the contents of any tank that might begin to leak into an empty reserve tank.

At Windscale ten of these tanks are in use holding a total of about 770 cubic metres of the liquid (1978 figures). This represents virtually all the fission-product wastes accumulated in the United Kingdom's entire civil and military nuclear programmes. In addition, a much smaller amount is held in similar conditions at Dounreay. However, the tanks and cooling systems need continual supervision and must eventually require replacement, and it is generally agreed among the nations that it would be more satisfactory in the long term if all such wastes were to be converted to a solid form. They would then require less supervision and would present an even smaller risk of reaching the environment.

With this in mind Harwell started work in the early 1960s on a process for turning the liquid wastes into glass. This can be done by feeding the liquid, together with glass-forming materials such as silica, borax and soda, into a stainless-steel cylinder heated electrically to around 900°C. Water and acid are driven off and the fission products, now mostly in the form of oxides, react with the added oxides to form molten borosilicate glass which gradually builds up to fill the cylinder, which is then capped and welded. Glasses have been developed that have great resistance to heat and radiation and, above all, to leaching of active materials by water. BNFL and the UKAEA together are developing the process to the full industrial scale required by an expanding programme of nuclear power. By about 1990 BNFL expects to embark on the routine conversion of existing stocks of liquid waste into glass. The blocks are expected to be in the form of stainless-steel-clad cylinders, about 3 metres long by 50 centimetres diameter. Ten of these blocks will contain all the fission product wastes arising from the generation of a million kilowatts of electricity for a year. About 3,500 blocks altogether would provide for all the high-level wastes arising from a nuclear power programme building up to reach 40 million kilowatts by the year 2000. Engineered underwater storage and cooling of the blocks would occupy only about the same area as two football pitches or a 300 metre run of motorway; then, when the rate of heat generation became less important — say in twenty to fifty years — the blocks would be disposed of away from man's environment and with no intention that they should ever be retrieved. Disposal would be by burial, either underground in suitable rock formations, or on or beneath the deep ocean-bed. Here the blocks would remain and in some 500 years the toxic potential of the radioactive fission products in them would have dropped by a factor of about a thousand. In 1,000 years the only activity remaining in them would be almost entirely from the actinides (curium, americium and a little plutonium). It would be comparable to the activity of the minerals from which the fuel was originally derived.

The disposal sites would be carefully chosen on grounds of geological stability and absence of earthquake and volcanic activity, and the rock itself must show good resistance to the action of heat, radiation and chemical attack, and the ability to conduct heat away from the glass blocks. The only path for a return of radio-active material from the disposal site to man's food-chain would be through the action of water, so it will be very important to see that the rock is free from water or fissures through which water might move. Even so, and notwithstanding the protection that the stainless-steel containers will provide for several centuries at least, the main safeguard will still be the retentive properties of the glass itself and its resistance to the leaching action of water, and this will be supplemented by the capacity of the rock (or of the back-filling material) to absorb any released chemicals. Research teams of geologists and nuclear experts are working in several countries, including Britain, to identify suitable rock formations. Granite, clay and salt look to be the most promising. This pre-liminary work will be followed by field tests, probably lasting for ten years or more, then a demonstration facility, before full-scale repositories are brought into use early next century. In Britain such a repository, if it is in hard rock and built to accept the 3,500

Fig. 36   *Radioactive waste management: If all the electricity — domestic and industrial — that one man used in his lifetime were generated by nuclear power, this would equal the total amount of glassified waste that would result.*

blocks referred to above, would occupy a volume of rock measuring about 150 m x 400 m x 150 m and located at least 300 m below the surface of the ground. This would lead to the production of about half-a-million cubic metres of mining spoil for disposal or landscaping. It would be the only spoil-ground of any size arising from forty years' operation of the entire nuclear industry of the United Kingdom.

The high-level waste problems are being tackled on broadly similar lines in other countries, notably in France, where an industrial glass-making plant is already in operation, and in Sweden, West Germany and the United States. Britain's work on assessment of hard rock-types forms part of the European Economic Community's research programme, and she is also working in association with the United States and Canada on some aspects of deep ocean-bed disposal — a course which would have to be subject to international approval. Some deep ocean sediments have lain undisturbed for many millions of years and could provide very safe burial grounds in which the resistance of the glass to leaching, rather than the presence of a stainless-steel covering, would be the major barrier to dispersal. But ultimate protection against the spread of long-lived radioactivity would be the slow migration and large dilution in the ocean itself, with absorption in the surrounding sediment as an additional barrier. Whichever method is chosen, the cost of vitrification and disposal by burial appears unlikely to add more than a very small percentage to the cost of nuclear-generated electricity.

One possible alternative to the glassification process would be to incorporate the fission products in a kind of synthetic rock, in which they would be held firmly at the boundaries between the crystal grains, to form a structure analogous to many minerals that are known to have remained stable for very long geological periods. This, its proponents claim, would provide a more stable structure than glass. Research proceeds on both routes.

9

SAFETY

Safety standards

No human activity is entirely without danger and the nuclear industry makes no claim to be an exception. However, it has a remarkably good safety record and in one respect at least it is unique among industries: in practical terms no nuclear installation is likely to be built or operated anywhere in the world unless a very high degree of safety has been designed and built into its equipment, its methods or work and the management of its wastes. According to UK Government statistics, it is the safest of the energy-producing industries and working in it is one of the safest of all industrial occupations. As far as its effects on the public are concerned the industry's record is equally good, and in Britain (and indeed the whole Western world) no member of the public has been killed as a result of a radiation accident at a nuclear power station.

Nevertheless, nuclear generation has certain very specific dangers to be guarded against: it deals with amounts of energy that are potentially devastating, and it gives rise to radiations that can be fatal. Let us look squarely at these hazards and at how they are being tackled.

Energy releases

Nuclear power exploits the force that holds the atomic nucleus together, thus forming the most powerful source of energy available to man. A few kilograms of fissioning uranium or plutonium can yield the power of thousands of tons of high explosive: a nuclear reactor contains many times this amount. Clearly such a concentrated source of power must be treated with the greatest respect and, although the conditions for a nuclear explosion are hard to achieve even by design, the energy is there, and no risk must be run of releasing it inadvertently. But the fear that a nuclear power plant itself might explode like an atomic bomb is without foundation. The conditions in a reactor are the very opposite of those needed in a bomb, and no serious-minded objectors to nuclear power regard it as a risk.

However, energy releases that fall far short of atom-bomb levels can still be very dangerous. For example criticality incidents, when enough fissile material accidently collects together in a suitable conformation for a chain reaction to start up momentarily, can cause bursts of radiation strong enough to kill anyone who happened to be nearby. But with proper design and operating rules, such incidents need never happen, and indeed they have proved to be extremely rare: one such incident, leading to two deaths, was reported in the USA during the wartime work on the bomb, but none involving loss of life has been reported from the peacetime nuclear industries in any part of the Western world.

Similarly, if a nuclear reactor were to be mismanaged or to go out of control a chain reaction could grow more quickly than it ought to, leading to the release, perhaps suddenly, of too much energy. Again, such happenings have been very rare indeed: the only such fatal incident in the Western world took place at Idaho in the United States in 1961 in a small experimental PWR. One of the maintenance staff contravened instructions and moved a control rod in such a way that the reactor went momentarily critical. The resulting burst of energy caused a steam explosion which killed the three men working in the reactor building. The events leading up to the accident were thoroughly investigated and the findings given wide and detailed publicity, so that everything possible could be learned from it by reactor designers and operators the world over.

Radiation and its effects

Preventing the release of fission energy on a large or small scale is not, however, the most troublesome problem of the nuclear industry: radiation is far more significant. It is present at all stages in the nuclear fuel cycle from mining and fuel preparation to reactor operation, fuel reprocessing and waste management, and its control calls for constant vigilance. We saw in earlier chapters the reasons for its pervading presence, and how the measures taken to control it affect the simplest operations, giving them, to the outsider, an aura of unfamiliarity and even fear. To the people on the job, however, they are just an additional, though acceptable, complication. Most people working in the industry have no fears of radiation or of the machines and materials that produce it: they know that the subject is well understood and its true dangers assessed, and they also know from experience that plant designs and operating rules, supported by protective and monitoring systems, provide an excellent assurance for their continued safety.

In the pioneering days of the early 1900s there was no reason to suppose that the invisible radiations given out by radioactive substances were going to prove particularly dangerous, but through bitter experience experimenters soon found themselves forced to take more care and eventually to make radiological safety a subject of study in its own right. Later, the development of medical applications of radiation, particularly X-ray examinations and the treatment of cancer with radium, led to the introduction of a much more thoughtful approach to safety questions. No longer was it just a matter of a few research workers trying to ensure their own and their colleagues' safety without too much diversion of effort: it was fast becoming a matter in which hospital departments had to meet their responsibilities both to the patients themselves and to the medical and nursing staff and others who might be exposed to the radiations as part of their working routine.

In 1928 the world's medical profession, acting through its international specialist body, the Congress of Radiology, set up the International Commission on Radiological Protection (ICRP). This became a permanent watchdog charged with keeping abreast of current knowledge and making recommendations on safety measures. ICRP concerned itself most particularly with determining and keeping under review the limits of radiation dosages to which people working with radiation or radioactive substances might be exposed without unacceptable risk of harm. Corresponding limits were also worked out for exposure of members of the general public. ICRP soon became accepted as the world authority on the subject and after half a century of work it continues to be so regarded. Its recommendations are accepted throughout the world as the best foundation on which to build radiological safety legislation and codes of practice.

In setting radiation dose limits it is not the high doses that have turned out to be the most important: their effects are well known and are not difficult to assess and guard against, and they are only likely to arise as a result either of serious defects in plant design or of accidents or grave human error. Much more important are repeated or long-continued exposures to levels of radiation too low to cause direct bodily harm. The accumulated effect of these may be an increased risk of an exposed person developing cancer later in life, or of his descendants being born with some genetic abnormality (though there is very little evidence of the latter occurring in man). In individual cases neither of these effects can be attributed with any certainty to radiation exposure, still less to a particular source or occasion, because a great many cases arise from causes (natural or man-made) quite unconnected with radiation. At the same time, many people exposed to above-average radiation levels are unaffected, but spread over an exposed population the results may be statistically significant. As such they must clearly be minimised by keeping the radiation dose to the population very low. However, because of the natural 'background' radiation to which we are all exposed attempts to eliminate them totally would be pointless.

This background of radiation, against which all life, including Man, has evolved, arises partly from radioactive elements in the Earth's crust and in the air around it. Uranium, thorium and their daughter-products, and potassium 40 which is a universal constituent of our own bodies, are among the most important. 'Cosmic radiation' also comes in from the Sun and outer space, particularly at times of high sunspot activity; normally, however, we are shielded from most of it by the thickness of the atmosphere. Background radiation levels vary widely with place, with time and even with job — the annual dose to an Aberdonian office worker may be half as high again as that to a Londoner doing the same work, but if the Londoner happens to be a high-flying airline pilot the reverse might apply. While there is no evidence that natural radiation does anybody any harm its long-term effect on evolution through genetic change must remain a matter for speculation. It is against this background that we must look at any additional radiation exposures arising from the activities of the nuclear industry. At present the industry adds, mainly through waste disposal, the equivalent of only about one five-hundredth part of the average natural background dose received in the UK. This is far less than the local variations and well below the dose received from a single annual diagnostic chest X-ray. X-rays and other medical applications in fact form by far the biggest man-made

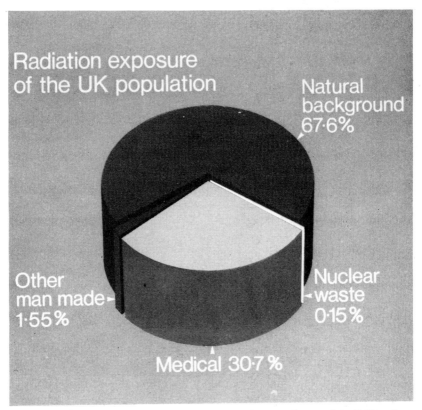

Fig. 37   *The sources of radiation contributing to the total radiation dose which the average member of the British public receives during a lifetime. The natural background, and therefore the total radiation received, will vary widely from place to place.*

contribution to our annual radiation dose (see fig. 37).

Most physical injuries are caused when an excess of energy is absorbed by the victim's tissues. What constitutes an excess in this context will depend upon the nature of the energy and on which tissues are affected by it. The radiations that we are concerned with injure by ionisation, that is to say by ejecting electrons from their positions in orbit around the nuclei of atoms. This interferes with the delicate chemical mechanisms on which all living processes depend. Man and the other higher mammals are the most complex of all organisms and are the most sensitive to this kind of damage, some organs being more sensitive than others. The human senses are of no direct help in detecting or measuring nuclear radiation, and this is one of the reasons why people tend to be frightened of it: they know it can harm them but they cannot tell whether it is present or not. But there are many kinds of instrument that can do the job, and even very low-level measurements present little practical difficulty: ionising radiations are in fact uniquely easy to detect and measure, far easier than, for example, chemical poisons at very much higher levels of harmfulness.

The most important measurement is of the energy absorbed per gram of tissue, with an allowance being made for the relative harmfulness of the different kinds of radiation — alpha, beta, gamma and X-rays, and neutrons. Exposures are measured in 'rems' which take into account the energy of the radiation and its effectiveness in causing harm to living tissue. Beta, gamma and X-rays are less harmful than alpha-particles (emitted within the body — outside they are harmless); these in turn are less harmful than fast neutrons. The intensity of the radiation at any one point is usually measured as the 'dose-rate', i.e. the dose that would be absorbed over a period of time at that point, expressed usually in millirems per hour (mrem/hr). (Under the new SI system of measurements the rem is superseded by the sievert, equal to 100 rems.)

Alpha-particles will only penetrate a fraction of a millimetre of tissue, but along their short paths they can cause very heavy ionisation. Since they cannot get through even the outer layers of the skin they do no harm externally, but they can do considerable damage if emitted within the body by the decay of alpha-active materials that have been swallowed or inhaled. Those that are retained for a long time in the body because of their chemical nature, e.g. plutonium, are therefore the most dangerous. Beta-particles can get through the skin and up to a centimetre or so of tissue. The distance that they can travel, and the intensity of ionisation along their paths, depends on their energy (measured in electron-volts, eV), which is related to their speed. The same also applies to alpha-particles. Gamma-rays and X-rays have no such limit to their penetration but they are progressively weakened in intensity ('brightness') as they give up their energy, until they are no longer detectable. They too cause ionisation in the tissues along their track, the amount depending upon the energy as well as the intensity of the radiation. Their energies are related to the frequencies of their vibrations, not to the speed of their movement, which is that of light. Neutrons, being uncharged, do not cause ionisation directly, but fast neutrons are very damaging both because they are very penetrating and because by knocking other atomic particles (protons in particular) violently out of place, they can indirectly cause very heavy ionisation deep in the body.

Starting with the highest doses (which would only arise in severe accident conditions), a dose of 500 to 1,000 rems would kill a person within a few weeks, with certainty. A single whole-body dose of 250 to 350 rems could be expected to kill about half the people exposed to it. These high doses cause general 'radiation sickness', with reduction in natural resistance to infection, internal bleeding, diarrhoea, vomiting, convulsions, and in extreme cases death. If enough sound tissue is left the body will make every effort to carry out its own repairs and the victim may recover. Doses of 50 rems and above to a single organ can affect its working adversely for a time because of the proportion of cells killed. (This effect is the one that is used for the benefit of patients undergoing radiation therapy, where very localised heavy doses are used to kill the cancerous cells in a tumour.) Below about 10 rems no directly adverse effects have been demonstrated among adults, but the foetus appears to be more susceptible, and doses of 1 to 2 rems can increase the risk of early leukaemia. For this reason ante-natal X-rays examinations are kept to a minimum.

Going down to lower levels, the effects are statistical rather than on the individual. If a dose of one rem were given to each of a million people about one hundred of them might contract cancer as a result – but about 2,000 times this number will, according to current expectations, die from cancer in any case. To put it in another way, a radiation worker who receives an annual occupational dose of one rem is running an added risk of about one in ten thousand – on top of the one-in-five risk that all of us already run – of dying from cancer as a result of each year's work. This makes his job as a radiation worker about as dangerous to him as smoking three cigarettes a week. As far as the genetic dose is concerned, if the same dose of one rem to each of a million people were spread over a generation of thirty years (this is the legal population dose limit in the United Kingdom) then, from extrapolation from animal experiments, about ten substantial genetic abnormalities might be expected per year. To put this into perspective, naturally occuring genetic diseases affect some 6 per cent of all babies: by comparison, the additional incidence due to the radiation would be so tiny as to be statistically imperceptible.

Determination of dose limits

The relation between radiation exposure and the likelihood of its causing long-term effects has been found out in part by studies of case histories of persons known to have been exposed to radiation, in part by experiments on other organisms and applying the results to man, and in part by epidemiological methods. Direct experiments on people are, of course, not possible, both for ethical reasons and because of the impossibility of isolating anyone from the all-pervading background radiation for a significant time. However, a lot of work has been done starting from the days of World War I and continuing to the present time. Examples include studies in the USA of a group of women whose wartime task was to paint instrument-dials with a luminous composition containing radium: some of them, through ignoring instructions not to use their lips or tongue to 'point' their brushes, took in measurable quantities of radium, and some of them developed cancer later on in life. A relation between their still-measurable internal radiation levels and the incidence of cancer was thus established, covering a period extending over more than fifty years and based on a statistically significant numbers of cases. Another study involved the case histories of patients who in the inter-war years were treated for the painful and crippling disease of ankylosing spondylitis by exposure to quite large doses of X-rays; this too resulted in some cancers later. (Radiation has long since been discontinued as the preferred treatment for this.) Examples of other groups studied by ICRP and other workers include survivors and their descendants from the atomic bomb attacks on Nagasaki and Hiroshima, survivors from the Japanese fishing vessel *Lucky Dragon* which was caught in the fall-out from an atomic weapons test over the Pacific in 1954, and long-term studies of the health of people living in areas where the natural background radiation is, for geological or other reasons, very much higher (or lower) than the world average. The hospital services and the nuclear industries and other occupations in which detailed individual exposure records have been kept for more than two decades provide another and increasingly valuable source of information. NRPB is currently working on a comprehensive study of all radiation workers in the British nuclear industy, and as many ex-workers as they are able to trace.

Further useful information is becoming available from the Defence services too. For example, a study has recently been done at the nuclear submarine depot at Rosyth in Scotland. About 200 dockyard workers exposed to 'occupational' levels of radiation, and therefore subject to regular medical examinations including blood tests, have been under observation by a Medical Research Council team for ten years. A relationship has been clearly established between radiation exposure at these low levels (5 rems per year is the annual maximum occupational dose) and minor changes in some white blood cells. While there is no indication that the changes (which also occur naturally) are in the least degree harmful the tests nevertheless indicate that even at these very low levels radiation does cause changes in cells. This supports the widely-held view that there is no 'threshold' of exposure below which radiation has no effects at all, but it does not imply that there is no threshold for effects that are harmful. It also confirms the value of blood examinations as providing a biological monitoring system that can usefully supplement routine radiation dosimetry, especially in assessing accidental or suspected over-exposures.

In spite of all these studies there is still little firm evidence about the effects of very low doses of radiation, so the ICRP takes the conservative view that all additional exposure above the natural background should be kept to a practicable minimum and that no increase in exposure should be permitted without some compensating advantage to the community or the individual. They have made recommendations on the limits of exposure for radiation workers, that is to say people whose job is with radioactive substances or ionising radiations, whether in industry, medicine or research, and who are therefore subject to close medical check-ups.

The picture is a complex one because some parts of the body are much more easily damaged, or their functions are more likely to be impaired by radiation than others. The gonads, which are the organs responsible for the continuance of the human race, are considered to merit the greatest protection and therefore to be subject to the lowest limit of dose. This is not because they themselves suffer damage, or because sexual function is impaired, but because of possible damage to the genetic coding carried by the germ-cells themselves, leading to adverse effects in succeeding generations. Bone-marrow is another sensitive organ because radiation can interfere with its function of producing blood cells. The cornea of the eye is itself liable to be directly damaged by radiation, which can cause cataract.

The maximum whole-body dose recommended for members of the general public is set by ICRP at one-tenth of the whole-body dose for radiation workers. This is because the latter are fewer in number and are under regular medical surveillance. In Britain the limiting levels of exposure recommended by ICRP have been endorsed by the Medical Research Council (MRC) and by the NRPB. These levels are interpreted by NRPB in terms of detailed limiting figures for particular radioactive isotopes in the environment and for the levels of exposure of workers and the public to ionising radiations arising from nuclear installations. It is on these limits of radiation exposure that much of our nuclear safety legislation is based and on which nuclear site licences and waste discharge authorisations are granted.

Radiation releases

There have, of course, been accidents involving the release of radioactivity, notwithstanding the precautions and regulations. In the Western world the first serious accident was the fire at Windscale in 1957. During a routine maintenance operation in which the graphite core had to be heated up, local overheating occurred and some of the fuel and graphite caught fire. The reactor was, by modern standards, quite small (about 150 MW) and its sole function was to produce plutonium. Not being a power station it had no pressure vessel and the core was open to the air, in the sense that its waste heat was removed by drawing filtered air through the reactor core, and out through further filters situated in the ventilator stack, to the atmosphere. In this incident the solid fission products liberated by the fire were all successfully retained on the filters but iodine 131, in the form of a vapour, escaped to the air. It was estimated that some 20,000 curies of iodine escaped, of which a large proportion was brought down by rain and deposited on agricultural and grazing land down-wind of the reactor. Some of the iodine was taken up by cattle grazing in the area and transferred to their milk, so that any person drinking local milk would have been liable to take in excessive iodine 131 and accumulate it in his thyroid gland. It was considered that babies fed wholly on cow's milk from this area might be at an increased risk of contracting thyroid cancer later in life, so it was all impounded for several weeks after the accident. By then the normal radioactive decay of the iodine made it safe to start using the local milk again. A subsequent report by the Medical Research Council stated that no injuries had resulted, or were likely to result, from this substantial release of radioactivity.

The Windscale accident led to considerable strengthening of safety measures, including the institution within the UK Atomic Energy Authority of an independent Health and Safety Branch. Another development was the setting-up at a number of major nuclear establishments, including Windscale and the Generating Board's nuclear power stations, of Local Liaison Committees (LLCs) to keep the public authorities in the area aware of activities on the site, and of precautions against accidents, especially those that might lead to releases of radioactivity, and to co-ordinate arrangements for the safety of local populations in such an event. Another specific measure that owes its origin, at least in part, to the Windscale accident and which bears on the activities of the LLCs, is the provision at nuclear power stations of stocks of iodide tablets for distribution to members of the local population in the event of an actual or threatened release of radioactive iodine. By saturating the thyroid gland with inactive iodine, the uptake of any radioactive iodine would be largely prevented.

Another major nuclear mishap, about which there is still much uncertainty, appears to have taken place in 1958 in the southern Ural Mountains region of the Soviet Union. Although nothing has yet been officially stated by the Soviet authorities, contemporary and later reports on radiation levels, and observations by some of the few travellers who have visited or passed through the area, combine to indicate that a major destructive event of some kind took place, leading to a large release of radioactivity and the long-term evacuation of a wide area. It is not clear whether this arose from an accident to a nuclear reactor or at some point in the fuel reprocessing cycle, including waste storage. What does seem clear, however, is that it took place at a military rather than a civil establishment: it is known that there are military plants in the area. The persistent lack of official information supports this view, and has also prevented any useful lessons being learned from the incident by the world's nuclear industry: here it contrasts sharply with Windscale and with the Three Mile Island accident described below.

Overshadowing all previous nuclear accidents, however, was the one at Three Mile Island, Harrisburg, Pennsylvania. At four o'clock in the morning of 28 March 1979, workers in the 960 megawatt PWR power station were alerted by a blockage occurring in part of the boiler feedwater purification system — a not very unusual fault in any power station, and one well catered for in the design. Automatic safety measures came into operation immediately and correctly: the feedwater pumps stopped, the turbine stopped and reactor power output dropped to about 15 per cent of normal. Emergency feedwater pumps came into operation, and it was here that the trouble started: no feedwater could get through because wrongly, and unknown to the operators, the pump valves were shut, so the boilers quickly boiled dry. This allowed the reactor water temperature and pressure to rise, and safety valves opened (correctly) to relieve the pressure. In about eight seconds the reactor shut itself down automatically, as it was designed to do, and from that moment nuclear fission ceased and the only heat generated was due to radioactive decay of the fission products in the fuel.

But the troubles were not over. One of the reactor pressure relief valves failed to close properly and there was confusion in the operation of the emergency core cooling system. Enough water escaped to allow some of the fuel to dry out and overheat, resulting in damage to the zirconium cladding, and liberating radioactive fission products which escaped with the water. By the time the pressure relief valve had been closed and cooling water supply restored much damage had been done, and a bubble of gas had formed in the reactor which prevented proper cooling. This contained some hydrogen (thought to be a possible explosion hazard — erroneously, as it turned out) from the zirconium/water reaction, and it proved very difficult to clear. Meanwhile, the radioactive water from the reactor had filled and overflowed the sump in the reactor containment building, and was being pumped automatically into a neighbouring building, from which some of the radioactive gases in it, including krypton 85, escaped into the atmosphere. The water was later pumped back to the reactor building and the radioactive release stopped.

Within hours the heat of radioactive decay had eased off, but it was several days before it became clear that the emergency — if there had ever been one — was at an end, and a month before natural-circulation cooling was established. Although partial evacuation of the area had been ordered as a (perhaps over-hasty) precautionary measure, it appears that the release of radioactivity was, in fact, very small and is likely to have hurt nobody, and its effect on cancer statistics in the years to come will probably be so small that it will be undetectable.

Although the Three Mile Island accident may prove to have been a setback to nuclear power, nevertheless a great deal will undoubtedly be learned from it that will help to improve the safety of nuclear power generation. It appears that there were shortcomings in the reactor instrumentation and in the regulatory procedures, also apparently in operator training. However, the physical safeguards worked, and there was no major outside release of radioactivity in spite of extensive damage to the fuel and the escape of reactor water. There was no danger whatever of a nuclear explosion, and no evidence to suggest that PWR designs are intrinsically unsafe.

To minimise the risks of failures resulting in releases of radioactivity, all nuclear plant has built-in safety and warning devices, and automatic shut-down systems are fitted where appropriate. For a fault to result in a major accident leading to a release of radioactivity, the safety defences would need to be breached at each one of the several levels of control and of containment. One after the other, or simultaneously, there would have to be: failure of the multiple instrumentation systems to detect and control the incipient fault; failure of the automatic shut-down system; failure to prevent the fault developing to a dangerous level; failure of the fuel elements to retain the fission products; failure of the primary reactor containment; and finally, failure of the secondary containment. The experience at Three Mile Island will be added to that gathered elsewhere, and will go to make nuclear power safer. By pointing up any particular weaknesses in the lines of defence at that specific reactor, the lessons can be applied to other designs and other locations.

Under British law, before a start can be made on building a nuclear power station a site licence must be obtained from the Health and Safety Executive (HSE). This, if granted, will apply to a particular fully-detailed design for construction at a specified site. The Nuclear Installations Inspectorate (NII), acting for the HSE, studies the design very closely, looking in particular for all the things that could go wrong with the reactor, its fuel and its auxiliary systems, under any credible set of circumstances. Working from their own records and from a study of accidents and lesser failures that have happened in all branches of industry (not just nuclear), specialist scientists and engineers examine in detail the ways in which individual components and instruments, etc., specified in the design can fail, the likely frequency of such failures, and what they can lead to. All credible chains of events leading to possible releases of radioactivity are examined in detail and their likelihood assessed. Possible consequences of such releases in the reactor's specified environment in various credible circumstances of weather, time of day, etc., are assessed in terms of direct loss of life or later increase in cancer mortality. The estimated frequency with which releases will occur is compared with the expected seriousness of the results in each case. In this way a firm and realistic base is created upon which NII can decide whether to advise that a site licence should be granted or to call first for modifications to the design or operating procedures. It is obviously unrealistic to expect 'no accidents, ever', but this method, though very time-consuming, helps to ensure that the most likely accidents are the least serious, while the most serious ones are the least likely to happen.

The NII's requirements are strict on this point: the licensee must show that an accident large enough to put members of the public at risk (but without actually causing casualties) would be prevented by a succession of protective systems or barriers such that the chances of a fault occurring and not being checked would be no more than about one in a million per year of reactor operation.

10

ENERGY AND THE PLACE OF NUCLEAR POWER

Energy

Earth carries with it through space limited capital stocks of materials, none of which change significantly in total quantity over the ages. Man finds and uses many of these materials to satisfy his needs, but in doing so he neither destroys them nor creates more of them: he only changes the way in which their constituent parts are put together. The few dozen chemical elements of which nearly all Earth's materials are made can be thought of as an enduring capital asset, in theory infinitely recyclable.

As well as carrying stocks of materials, Earth carries its own capital stocks of energy. Some of this is hidden deep inside as heat, some is locked up in the nuclei of atoms, to be liberated in nuclear power stations or through radioactive decay. Some that has come in from the Sun in the past is temporarily stored up in the 'fossil fuels' — coal, oil and natural gas. Unlike the chemical elements, these stocks of energy are not recyclable: in terms of human history, once used, they are lost for ever, so if we use up for ourselves all that we can reach of the energy of Earth's fossil fuels and of its uranium, those who come after us will have to dig deeper or look elsewhere for energy, or they will have to face, by comparison with us, an energy-hungry way of life.

Most of Earth's energy supplies come in daily from the Sun, which warms the ground and the air, causing winds, rain and ocean currents. Some of it arrives as light and is absorbed in giving life to plants and trees. Most of it, however, is radiated back into space next night, to be replenished again (more or less) in the rhythmic cycle of days and nights, winters and summers. Until the Industrial Revolution, it was this constantly-replenished energy that supplied nearly all Man's needs, but with the advent of steam-power he turned to Earth's capital reserves of energy, coal in particular, then to oil and gas and later to uranium, and he has been using them up at an increasing rate ever since.

Besides being useful in itself for warmth and light and for doing work, energy has a unique property: man can use it to alter the way in which Earth's materials are put together, by changing one substance into another one that he wants more. For example, he learned long ago how to bake clay to make useful bricks or pots, using energy that came in from the Sun and has since been locked up in fresh or fossilised vegetation. Similar principles apply to many of the other materials that form the basis of our industrial civilisation — metals, heavy chemicals, fertilisers, cement, etc. Most of these are made by chemical or metallurgical processes requiring the use of energy. If an easily-won material, copper for example, becomes scarce, a more abundant substitute, aluminium, may be used instead, provided that enough additional energy is made available — at an acceptable price — for smelting it. Similarly, synthetic oil for use as a premium transport fuel can be made, at a price, from a less convenient fuel such as coal, given enough additional energy for the conversion: part of the coal is used up to change the rest of it into a more convenient fuel. In terms of energy we pay particularly dearly for the convenience of using electricity: about 60 per cent of the coal's energy is used up in the power station to turn the other 40 per cent into electricity. Re-cycling of materials and their recovery from wastes often require additional energy, and in an energy-hungry world are likely to be neglected until the need for a material matches the extra cost of recovering it. The same applies to the prevention or clean-up of pollution, so an energy-hungry world can soon become a dirty world as well as an uncomfortable one. Some shortages of materials, then, may be eased if there is an abundance of energy. But a shortage of energy itself, as well as bringing its own hardships, can lead to or aggravate shortages of material.

Oil, gas and coal share one important property that is not related to their use as sources of energy: they are largely made up from carbon atoms linked together in chains or rings with hydrogen and other atoms attached, and this is just how the makers of plastics, textile fibres, drugs and numerous other chemicals need them. So whenever we burn fossil fuels, we gratuitously discard a non-renewable asset in the form of linked carbon atoms. We must keep this in mind in deciding how much of our fossil fuels to burn and how much to use as chemical feedstock. For instance, it is already arguable that natural gas is too valuable to burn, and yet we use increasing amounts of it for domestic heating and cooking, for which it is our cheapest fuel. Is this wise? Perhaps it ought to be priced on its usefulness rather than on its production cost.

Energy, then, is all-important to Man, especially if he wishes, as most of us do, to have a reasonably clean, comfortable and orderly life based on an abundance of goods and services. Yet thanks to our own complacency and that of our forbears (and world trends that are beyond our control), we now have to face the real prospect of an energy shortage. Although detailed forecasts differ, sometimes very widely, it is generally accepted that, if we go on as we have been doing, Britain's and the world's capital resources of oil and gas will be starting to run low around (or even before) the turn of this century, followed by a shortage of uranium early in the 2000s. Coal alone is expected to remain available in quantity for several centuries. Optimists may put these dates a bit further ahead, but few will be found who deny the principle.

Consequences of an energy shortage

Before looking at remedies let us first consider what would be the result of an energy shortage. Fuels will not suddenly run out but, as we can see only too clearly even now, they and the power they produce will progressively get more expensive in relation to what we pay for other goods and services. This increase will be reflected in other cost increases, probably led by transport, energy-hungry manufacturing processes, and goods based on the petro-chemicals and plastics industries. Many metals too will become less easily available. There will be a general shortage of purchasing power and services, a decline in trade, dwindling exports with dearer imports, increased unemployment, and a general fall in living standards. The resulting cuts in social spending will hit first and hardest the old, the sick and the poor. Aid to the developing countries will diminish or disappear, through lack not of will but of capability. In a mainly rural society, much of this austerity might be made more bearable — indeed rather attractive to some people in a grim sort of way — through the pursuit of individual or local schemes of self-sufficiency in food, energy and domestic requirements. However, in an industrialised urban society such as we have in Britain, there is literally very little room for significant self-help of this kind. Populations of big towns could be faced with disruption of essential services and supplies, leading to real hardship and even to a possible breakdown of the structure of society.

Energy shortage, much more than shortages of materials (for which substitutes may be found), transcends national frontiers. Multiplied through the nations, a situation could develop of extreme international tension between 'haves' and 'have-nots', leading to very real danger of war. The countries of the Third World are particularly vulnerable to a falling-off of world trade or a major rise in the real costs of energy. To support their rapidly growing populations they must adopt new techniques of food production, and this means importing large amounts of agricultural chemicals and machinery, which can only be paid for by exports. They must build up industries to produce these exports and to provide work for the growing adult population. All these activities depend on stable world markets and reasonably low-cost energy, particularly oil for farm and factory.

Conservation and resources

How can we meet the situation, especially here in Britain? Conservation of resources must come first: we must try to cut out direct waste of energy wherever we find it, then introduce practical energy-saving schemes where these can make an economic contribution both in the short term and on a small scale – e.g. house insulation – and in the long term and on a larger scale, in the form of combined municipal heat and power schemes, for example. Yet not all such schemes, however enthusiastically conceived and advocated, are economic: the total energy costs of a scheme must be taken into account as well as all the other costs, and these together must be balanced against the savings expected over its lifetime. The disruption caused by its introduction must also be recognised.

We can learn to be more discriminating in using our limited capital assets of fossil fuel, restricting them to those uses for which they are best fitted, either as a premium energy source (e.g. oil for transport) or as feed material for the manufacturing industry, rather than expending them wastefully merely to produce heat or electric power. We can cut down on unnecessary or frivolous uses of energy, voluntarily or by legislation, rationing or price policy. We can look critically at the economics of private cars (and at our own driving habits) versus public transport, and at road versus rail or waterway for the movement of goods. But conservation alone cannot meet the gap unless we are prepared to accept restrictions of an austerity that could be reminiscent of wartime, and unless all the technological advances that would be needed proved successful beyond the normal pattern of experience. Let us then look briefly at each of our main energy sources in turn — fossil fuels, renewable resources, and finally nuclear energy.

Here in Britain coal is without any doubt still our major energy resource: we have assured stocks to last us, at the present rate of consumption, for two or three centuries, and new discoveries are still being made. Most of our coal is deep-mined but some is from open pits. Technical improvements are being made in mechanical mining and in combustion methods to give greater efficiency in use, with less pollution and less personal hardship to miners. Ways are being found to liquify coal to replace the fast-diminishing petroleum oils as fuel for transport and as chemical feed-stocks. Nevertheless, it is arguable that coal, like oil and gas, may soon become too precious to burn. Winning coal gets no easier or cheaper and, in spite of much improvement, it is still a dangerous occupation. Whatever the reserves below the ground, there is a limit to the rate at which they can be extracted, depending on the availability of capital and the productivity of labour — which certainly gets no easier to recruit. In any case, significant increases in total output can only take place by degrees; it takes ten years to open up a new pit and, if the coal is to be burned to make electricity, about the same time to build and commission a new power station. Any hurried expansion of the industry could lead to a temptation to relax standards of safety and of environmental impact, in an industry that already has enough of these problems.

Some authorities believe that continued very large-scale use of coal and other hydrocarbon fuels could lead to the so-called 'greenhouse effect': an increased concentration of carbon dioxide in the air might restrict the re-radiation of heat from the earth's surface at night, leading to an increase over the years in mean air temperature. This could eventually lead to large-scale melting of the polar ice-caps, with possible flooding of huge tracts of low-lying land. It is not a certainty that this will happen, but it is a possibility that should not be ignored. Growing pollution by sulphur dioxide should not be ignored either: it is a fact, not just a possibility. Finally, coal is the most expensive fuel to transport, and is very often to be found a long way from the users' market.

Ten years ago the United Kingdom had virtually no known oil resources of her own: we now know that under the seas around the country we have some of the largest proven oil reserves of any country in the world, and fresh discoveries continue to be made. However, we are already tapping our known resources, or preparing to tap them, at a rate which on most present estimates will lead to peak production in the 1980s or 1990s followed by a gradual decline, moving towards scarcity in the early 2000s. We must consider whether such rapid exploitation, largely by burning in power stations and space-heating systems, is really the best course to pursue. As we have seen, oil and gas are ideal feed-stocks for important manufacturing industries, and burning them (except for premium uses, e.g. transport) is wasteful. It could also add to the 'greenhouse effect', if this proves to be a reality.

Similar considerations apply to natural gas, another bonus discovery made a few years before North Sea oil was found. We are exploiting it, again very largely for burning in heating systems, as fast as we can win it. Already one-third has been used, and at present rates its production will reach a plateau in the 1980s and start to decline before the turn of the century. Again we should question the wisdom of burning it indiscriminately, losing all its inherent chemical value as well as its special convenience as fuel, e.g. for domestic cooking and some chemical processes. The flaring-off of gas from oil wells, whatever its technical justification, is an affront in terms of resource conservation.

Renewable energy sources

Let us now turn to the so-called 'benign' and 'renewable' energy sources and see how these can help Britain in particular. In the UK hydro-electric power is already being used about as fully as it can in the situations that favour it, that is the mountainous areas of Scotland and Wales, and there is little scope for further expansion. 'Pumped storage' schemes, where a hydro-electric scheme is kept 'topped up' by the pumping power of a large base-load station, often a nuclear one, allow power generation to be spread evenly over the day and night cycle. By meeting the increased demand of day-time, and especially morning and evening peak periods, by hydro-electricity, these schemes economise in the production and use of energy rather than actually adding to the supply.

The twice-daily rise and fall of tides, caused by the Earth's rotation upon its own axis and the orbiting of the moon around it, was for many centuries a valued source of energy in the United

Kingdom and elsewhere, particularly for grinding corn — there were 28 tide-mills on the River Thames at one time — but it did not prove competitive with modern methods and has not survived. However, interest is now alive again and studies have been commissioned on schemes for using the very large Severn Estuary tides, but the cost would be very high and there would be many social and environmental problems to be solved, for a limited return in electric power generation.

We can use some of the heat, light and other radiation that comes to us directly from the Sun. In the United Kingdom the most promising major application seems to lie in the heating of buildings or water, in schemes ranging from single roof-panel collectors to boost a domestic heating system, to commercial or community buildings designed from the start to make the fullest use of solar heat. Research and development effort is being allocated on an increasing scale from public funds, and commercial exploitation of solar heating technology is growing.

The production of electricity by conversion of sunlight directly into electric current by arrays of photo-electric 'solar cells' is unlikely to make any significant contribution to our energy supplies by the turn of the century, if ever. The amount of sunlight in United Kingdom latitudes is small and ill-matched to electricity requirements, while the technology of conversion is, and looks like remaining, very expensive. Efforts are being made to harness some of the natural processes of solar energy storage, e.g. by growing appropriate crops which can themselves be used as an energy source, either by burning (in the way that wood has been used through the ages), or through biochemical processes such as fermentation: alcohol has possibilities as a good fuel for transport, and methane gas burns well.

Public and privately funded research on various ways of using solar energy is continuing on an increasing scale, based on judgements of the prospects of technical and commercial success. In sunnier countries, of course, the prospects are very much better than here in Britain, but there is a very long way to go before the Sun catches up with the atom as a useful source of power in any of the industrialised countries.

The wind, Man's most ancient source of mechanical power, shows promise of providing a fresh contribution to Britain's energy needs, through up-to-date designs of wind-driven generators sited in exposed areas. However, there are difficulties in matching supplies to demand, both in location and in timing, as well as difficulties in mechanical design and construction. An array of wind-driven generators having the output of a single 2,000 MW power station would have an enormous impact on the environment. Several hundred large and very strongly built rotors would have to be put up in prominent places, with collecting power lines. These would be in areas which, if the scheme were to work well, would probably also be areas of special beauty or appeal, such as the coast and hills of northern and western Britain, or the flat lands and rolling Downs of the south and east. Generation rates would fluctuate wildly and some way of storing the energy would be needed, locally or centrally.

There is vast energy in the waves around our shores, and its seasonal arrival to some extent matches demand. Several promising schemes for harnessing this energy to electricity production are in hand in the United Kingdom, under Government sponsorship. There are major problems of design, construction and maitenance to be overcome, and of transmitting the energy from the generating installations on the open sea to the distant land-based areas where the demand for power will be located.

Energy from the hot interior of the Earth comes to, or very near, the surface in a few localities, e.g. in Iceland, New Zealand, Italy, Japan and parts of North America. In Italy some of the heat drives power stations supplying the country's railway system, and there are district heating schemes in operation in France and Iceland. In parts of southern England there are possibilities of tapping this heat either directly as hot water or by pumping water down to the hot rocks and getting steam out, but its potential contribution to our energy supplies is very small. Geothermal energy ought strictly to be regarded as a capital, not a renewable resource: at any given place it can eventually all be used up.

All the renewable sources of energy taken together, assuming that each one of them were to be successfully harnessed, could in the Department of Energy's estimation produce only the equivalent of 10 million tonnes of coal, or about 4 per cent of Britain's primary energy needs by the turn of the century. There are, of course, other estimates of the contribution in terms of megawatts which are dependent on judgements of the prospect of successful development in harnessing each of the various sources; in terms of per cent contribution to the supply they depend on assessments of future energy demands. Forecast figures naturally range very widely, but it is noticeable (and to be expected) that estimates by Government or industrial bodies that carry responsibility for meeting the energy demands of the future are generally a good deal less sanguine than those of academic or voluntary organisations and of individuals who do not carry any such responsibilities. Even to get reliable forecasts a great deal more research must be done.

Nuclear power

In contrast, nuclear power is very well established as a technique for large-scale electricity generation, for which it has proved economic, safe and clean. It is capable of very substantial expansion without the need for extensive research and development, and it can cushion the effect of the depletion of the world's oil and gas reserves during the years when nuclear fusion and the renewable energy resources are undergoing development towards making major contributions to the world's energy supplies, hopefully in the early 2000s.

Of all the fuels in Earth's capital stocks, only uranium is an energy source and nothing else: the element has no other significant commercial uses, but it is the primary fuel of all the world's nuclear power stations. Reserves of uranium are fairly widely distributed in the Western world, but in the United Kingdom we have no substantial uranium deposits (although small quantities occur in and around the granite areas of Scotland and its islands, and in south-west England). Most of the uranium burnt in Britain's nuclear power stations has come from South and South-West Africa, with some from Australia, USA and Canada. These are the Western world's main supply areas.

The effective size of the world's uranium reserves (like that of all mineral resources, fuel and others) depends less on the total amount actually in existence in the Earth's crust (or even in the accessible parts of it) than on the effort we are prepared to exert in getting it out — that is to say, how much we are prepared to pay to find and win it: it is a question of economics as well as geology. In other words, it is only possible to put a figure on the amount available if we also state the maximum economic cost of extraction at current levels of technological development. Figures for resources can then be worked out using geological information about the location, nature and richness of well-surveyed deposits. Less firm estimates of reasonably assured reserves can be made, based on the apparent extent and richness of less well-prospected deposits.

In addition to the land reserves, the oceans of the world contain a further very large quantity of uranium. Ways of getting this out have been studied, but at present they are, and look like remaining, prohibitively expensive, so the oceans' uranium resources will probably remain unexploited for a very long time. Nevertheless, they exist.

As can be expected, estimates by different authorities of uranium reserves in the West vary widely, even at given maximum prices. Estimates of consumption rates also vary to an extent depending mainly on assumptions about economic growth and the size and rate of growth of the world's nuclear power programmes and, looking further ahead, on the expected ratio of thermal to fast reactor construction.

All estimates and assumptions about the future are, of course, fallible and experience has shown that energy forecasts – and nuclear energy forecasts in particular – are likely to be wide of the mark. However, most forecasts by British and overseas authorities indicate that economic resources of uranium throughout the Western world, if used to feed power programmes based on thermal reactors alone, are unlikely to stretch much beyond the first quarter of the next century. The same appears to be the case in the Eastern Bloc countries.

Fast reactors

By far the greatest contribution that nuclear power and the world's uranium resources can make to meeting the world's future energy requirements will be through fast reactors. These burn uranium fifty to sixty times more efficiently than the present-day thermal reactors, and are capable of carrying a major proportion of the world's power load for several centuries, even if energy demand were to rise far above current levels. In the United Kingdom alone, for example, there are sufficient reserves of once-used (depleted) uranium already stock-piled at our nuclear fuel factories to equal in energy-content the whole of our known and as yet unmined coal reserves. This could carry us through the whole of the next two centuries and probably well beyond. Having such large supplies of our own will help to insulate us against fluctuations in the world prices of other energy sources, including uranium itself.

11

OTHER USES OF
NUCLEAR ENERGY

Nuclear shipping – naval

We saw in Chapter 3 that the work on nuclear weapons led on to
the development of nuclear power by two separate routes. In
Britain, gas-cooled reactors of the kind used to make plutonium for
bombs were redesigned to produce heat for electric power generation, while in America reactors of a different type – pressurised
water reactors – were developed specifically for nuclear submarine
propulsion, and it was largely from these that the PWRs (and later
the BWRs) were developed for use in nuclear power stations.
Civil work would never have been pursued to fruition if it had not
had the earlier military work as a basis: the cost would have been
frighteningly high and the difficulties too great for anything but
military necessity to have overcome. Even more daunting, perhaps,
would have been the direct development of nuclear propulsion
for shipping.

A small nuclear reactor is just about the ideal source of power
for driving a submarine. It uses very little fuel: a single charge of
enriched uranium oxide fuel will keep the reactor going at up to
full power for hundreds of thousands of miles before the vessel
needs to return to port for a replenishment of the reactor core.
Because nuclear fuel burns by a self-sustaining fission reaction in
which no oxidant is needed, the reactor does not need to 'breathe'
in the sense that a diesel or gas-turbine engine must take in air and
release exhaust fumes, nor does it need to store up power in
batteries for underwater use. The nuclear reactor supplies power,
on and under the water, for as long as the fuel lasts, with enough
auxiliary power available to extract from seawater as much oxygen
and fresh water as the crew may require.

In 1954, the world's first nuclear submarine, the USS *Nautilus*,
was launched. Commissioned in the following year she proved an

unqualified success, and three years later she left her home port on a routine voyage and (almost incredibly) sailed under the whole North Polar ice-cap, passing beneath the Pole itself. Now more than 250 nuclear submarines, nearly all of them in the navies of the United States and Russia, with a few in the British and French navies, are in port or cruising beneath the sea. All of them, so far as is known, are powered by pressurised water reactors. USS *Nautilus,* after almost a quarter of a century of service, was taken out of commission early in 1979. She and her sister ships provide an unspoken testimonial to the safety and reliability of the PWR.

Submarines alone can take full advantage of the nuclear reactor's capacity to operate without 'breathing' or producing exhaust fumes, but surface vessels too can benefit from the ability to go for very long periods without refuelling, and this gives them an advantage of independence both from port facilities and from the services of refuelling tenders at sea. A nuclear aircraft carrier, for example, can be regarded as a semi-permanent mobile floating air-base operating independently of fuel and water supplies for many years at a stretch. The US Navy has eight such vessels afloat today, the largest of which, USS *Nimitz,* of 194,000 tons dead-weight and the world's biggest warship, is capable of cruising 800,000 miles — thirty times round the world — over a period of thirteen years without a reactor core replacement.

Nuclear merchant shipping

The ability to run for long periods on a single charge of fuel, at full power if necessary, was very soon seen as a tremendous advantage for an ice-breaker operating in arctic regions. In 1959 the Russians commissioned the world's first nuclear-powered ice-breaker, the *Lenin* (see fig. 38), powered by two nuclear reactors of a design using pressurised water in multiple pressure tubes (rather than in a single pressure vessel) and graphite as the moderator (see Chapter 6). The *Lenin* was an immediate success and was given the job of keeping the strategic ports of northern Russia open throughout the year. She was followed by the *Arktika* of 18,000 tons, with two 'conventional' pressurised water reactors driving four steam turbines with a combined output of 75,000 h.p. In August 1977 *Arktika* cut her way through to the North Pole, being the first surface ship to reach it. (The first ship to break surface at the Pole was the USS *Skate* in 1959, the year after

Fig. 38   *The Russian nuclear ice-breaker* Lenin *ploughing through heavy ice.* *(Courtesy of Novosti Press Agency)*

Nautilus had passed beneath it.) In 1977 *Arktika's* sister ship, *Sibir,* completed her sea trials.

In May 1978 the Canadian Government authorised the design of a 150,000 ton nuclear ice-breaker, more than twice as powerful as the *Lenin.* The ship will be Canadian-built but powered by an American pressurised water reactor. Its task will be to open up a large area of the ice-bound Arctic Ocean for exploitation of the very large amounts of oil and natural gas thought to be available in this region. The vessel is expected to be completed in 1985.

Nuclear naval vessels and ice-breakers have been a success from the start, but this cannot be said for nuclear merchant shipping. The first such vessel was the United States nuclear merchant ship *Savannah*, built at Government expense and commissioned in 1962 as part of President Eisenhower's 'Atoms for Peace' initiative. It was hoped to show the feasibility of nuclear propulsion for merchant ships and to pave the way for the world's shipping and harbour authorities to accept nuclear ships as a matter of routine. It was never intended that *Savannah* herself should be a money-making investment, or even be commercially self-supporting, and as things turned out her achievements in this direction were small: the world was not ready for her, many ports refused to admit her for fear of accident or radioactive contamination, and in 1978 she was withdrawn from service and 'moth-balled'.

Meanwhile, other major industrial countries studied the problems and prospects of nuclear ship propulsion. Britain, in spite of her successful experience with nuclear submarines, concluded in 1971 that although it might be technically feasible to use any one of several different reactor designs, no nuclear merchant vessel stood much chance of commercial success in the foreseeable future, unless oil prices, in real terms, were to increase greatly. In 1974 a fresh look was taken at the problem both technically and economically, taking into account the 1973 oil crisis. Again, no decision was taken, although this and more recent oil price rises made the economics of nuclear tankers appear more attractive — a conventional tanker may burn up to 10 per cent of its cargo in order to reach its destination. However, the slackening in world trade discouraged the prospect of building, although Britain almost certainly has the technical and industrial capability of building ships powered by either pressurised water reactors or modified Magnox-type reactors.

In 1977 a firm of supertanker operators negotiated an order with an American shipyard for three 600,000 ton nuclear oil-tankers intended to travel 25 per cent faster than conventional craft and to cut costs by a quarter. Each tanker aims to carry five million tons of crude oil per year.

In 1968 West Germany launched the *Otto Hahn*, a 16,870 ton mixed cargo carrier powered by a modified form of pressurised water reactor. During the ten years of its life it travelled 600,000 miles without incident, and gained entry permits into the ports of 22 different countries. During these years of operation the total amount of fuel consumed was 57 kilograms of uranium 235.

At the end of 1978, when further fuel replacements would have been required, the Federal Government took the decision to scrap the vessel: it could serve no further useful scientific purpose to balance the money that would need to be spent on it, and in any case it was the Government's opinion that nuclear merchant shipping would be uneconomic for the next twenty to thirty years.

Japan, a world leader in ship building, decided in the 1960s that an early start to the development of nuclear shipping was vitally important to her. To gain experience the 10,400 ton nuclear vessel *Mutsu* was built at the town of the same name and fitted with a 36 MW(e) pressurised water reactor of American design. Early in 1975 she was ready to set out for her first sea trials but found herself blockaded within her own home port by hostile fishermen, and here she remained for about two years. Her sponsors had reckoned without the very understandable fear of radioactive pollution that had so hurt the people of that country in the past. At last, under cover of a typhoon, *Mutsu* slipped out — only to run into more trouble. Because of a design defect in her reactor shielding it was found that levels of gamma-rays and neutrons were much higher in certain areas around the reactor than they should have been (although they were nowhere actually dangerous at the reactor powers achieved at the time). Substantial alterations, however, would have to be made to the biological shielding before the reactor could be brought safely up to full power and the trials continued. It took seven weeks to persuade the angry fishermen to allow the vessel back into Mutsu City harbour, and then only on condition that a new home port be found for her within a year. At last in the late summer of 1978 she was accepted at a shipyard in Sasebo, where local unemployment was rife. Modifications are expected to take several years to complete.

In spite of this major setback — or perhaps because of it — plans to build a second ship have been debated, and with the experience gained from *Mutsu* Japan intends to enter the world market with low-cost reliable nuclear merchant ships to be built in batches of five to ten according to demand.

The major difficulties remaining in the way of widespread growth of nuclear ship propulsion are clearly organisational and human rather than technological: international safeguards and maritime law, liability for damage, crew training and insurance, and the broader issues of public acceptability, are some of the areas in which improvements are needed before nuclear merchant ships can come into their own. Of course, their operators also need the motivation of commercial or national gain, which is itself dependent upon the achievement of a satisfactory level of international trade.

Other nuclear transport

The use of nuclear power for land and air transport has often been suggested but it shows no signs of being adopted. The shielding necessary to make the reactor safe would be far too heavy and bulky to be worth carrying around, and the measures necessary to protect the operators and passengers, as well as the general public, both in normal running and in the event of accident, would be too difficult and too expensive to fulfil satisfactorily. It is in any case unlikely that the public would accept this extension of the application of nuclear power now or in the foreseeable future. The use of nuclear power in transport is much more likely to be indirect, either as an extension of conventional electric traction or in the production, by means of nuclear-generated electricity, of 'synthetic' fuels such as hydrogen, alcohol or hydrocarbons to be burnt in internal combustion engines.

Nuclear energy in space

Nuclear power has been proposed for the propulsion of space vehicles and for providing the power needed by satellites and lunar or planetary modules. Among engines that have been designed for this purpose is the American NERVA rocket which would use a small nuclear reactor to boil liquid hydrogen and eject it at very high speed through a rocket nozzle. It would be used for very long distance interplanetary journeys rather than for take-off from Earth. The same considerations of safety, however, apply at take-off and at re-entry (whether or not this is planned) and the project is hardly likely to be publicly or internationally acceptable in the foreseeable future.

The nearest practical approach to applying nuclear energy to space research is to use the heat of radioactive decay as a source of electricity for instrumentation on satellites and space probes. A few grams or kilograms of a suitable radio-isotope (strontium 90, plutonium 238 or uranium 235) are suitably enclosed so that the radiation energy is converted effectively to heat; this heats a bank of thermocouples which produce electricity. An assembly of this kind forms a long-lasting power source independent of solar energy collectors, by which the satellite can send messages back to Earth or carry out other allotted tasks. The radioactive source can be designed so that in the event of re-entry into the Earth's atmosphere (intentionally or otherwise) it either burns up and disintegrates completely or it survives intact and is recovered. However, this does not always happen according to plan: in 1977 a Russian satellite re-entered over Canada and broke up, but the radioactive source (uranium 235) was not completely dispersed and large pieces of radioactive hardware reached the ground, fortunately in a remote area – as did the very substantial (though not radioactive) debris of Skylab a few years later.

Other applications of radioactive decay energy

As a result of nuclear research and power generation, radioactive isotopes of about fifty different elements are now available commercially in a great range of physical or chemical forms and at economic prices. The uses to which they are put could form the subject of a separate book. It is appropriate to mention here some of those uses that, like satellite power sources, make direct use of the energy of radioactive decay. Microwave relays and radio transmitters for use in very remote or inaccessible places have been built, and navigational lights in which electricity produced in thermocouple devices is used to power a flashing beacon, or even a complete lighthouse, have been designed and operated experimentally with great success. Much simpler devices, based on the conversion of radiation energy directly to light by a phosphorescent material such as zinc sulphide, have been used for many years in, for example, the self-luminous signs and notices used in aircraft cabins or on telephone dials and (one of the earliest of all uses of radioactivity) on the luminous dials of clocks, watches and instruments.

There are a number of very important uses of radioactive decay energy in medicine. A recent one of these makes direct use of this energy (from plutonium 238) in the batteries of implanted heart pacemakers (see fig. 39): the principle is the same as in the satellite applications, etc., but the electricity is used to provide timed impulses to keep the heart beating when the body's own nervous system is not doing the job properly. Chemical

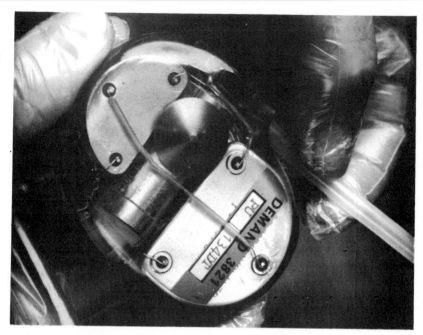

Fig. 39   *A heart pacemaker for implantation within the body. The battery, based on plutonium 238, has a design life of 20 years.*

batteries need much more frequent replacement surgery.

A much earlier medical use of radioactivity is in cancer therapy and related fields where the radiation, particularly highly-penetrating gamma-radiation, is used directly to attack and kill malignant cells, especially where these are deep in the body or otherwise not readily accessible to surgery. The same biological effect of radiation is used for bulk sterilisation of pre-packaged and disposable medical and surgical equipment — a development that has brought about a quiet revolution in clinical practice. Bulk gamma-irradiation is also being developed, largely under the aegis of the World Health Organisation, the International Atomic Energy Agency and the Food and Agriculture Organisation of the United Nations, as a means of prolonging the storage life of a number of different foodstuffs and for the elimination of insect and other pests. Here again, public acceptance of the process seems likely to be more difficult to achieve than its technical feasibility, which has been established for many years.

A smaller scale but widely practised application of energy

derived from radioisotopes is in the provision of portable sources of penetrating radiation for making radiographs — shadow pictures akin to X-ray photographs — of subjects ranging from engineering components and undersea oil pipes to the marble columns of the Parthenon in Athens.

12

LOOKING AHEAD

Some matters of special concern

There are a number of factors connected directly or indirectly with nuclear power which cause persistent concern among the public, and particularly among groups that can be described broadly as 'environmentalist'. It is convenient to collect these points together and look at them in turn and at the replies or reassurances given about them by the supporters of nuclear power. These replies extend the field of the argument to the consideration of the need for, and the alternatives to, nuclear power and to the existence and acceptance of other hazards. Eventually, they lead into fields of value judgements, politics and issues of right and wrong, and here each person must form his own views from his own interpretation of the information available to him. He must assess its reliability and examine the motivation of those providing it, and this may be no easy matter in a field as complex as nuclear power. Clear thinking is more important than deep knowledge. Let us look in turn at some of the arguments.

Atomic explosions

In 1945 the world saw only too clearly the devastation, death and disaster that even a 'small' atomic bomb could cause. Nuclear power reactors contain the same kind of material undergoing the same reaction of nuclear fission. Might not an accident or human mistake, or an act of sabotage, cause a nuclear reactor to explode in the same way? If it could, then surely, the argument runs, the risk cannot be worth taking and we should drop nuclear power altogether. This is understandable, but the entire nuclear industry supported by world scientific opinion is at one in affirming that an atom-bomb type of explosion by a nuclear power reactor is not merely very unlikely, it is impossible. The fissile material is far too scattered and diluted and largely of the wrong isotopic composition, and it has great quantities of non-fissile material among it. There is no means by which a sufficient quantity of fissile material could be brought and held together while the very rapid chain reaction spread through it — the material would disperse like unconfined gunpowder and the reaction would peter out without going beyond the stage of a criticality incident or a reactor power excursion.

The fear of an atom-bomb-like explosion in a power reactor is no longer regarded as a serious objection to nuclear power, even by those who oppose it on other grounds. Nevertheless, for many people the emotional impact of Hiroshima and Nagasaki and the later test explosions has been so disturbing that the mental image of a toadstool-shaped cloud is still conjured up by the very words 'nuclear' and 'atomic'.

Releases of radioactivity

If the nuclear industry's claim for such a very good safety record is well founded, then why, it is asked, do we so often read about leakages of radioactivity? If such incidents are trivial, why do newspapers report them? There seem to be two reasons for the attention given by the media to accidents with radioactivity: the first is the straightforward journalistic reason that it makes a good story which people will read. Radioactivity has been 'in the news' ever since it burst on the world at Hiroshima, and in this time of widespread environmentalist and humane concern it is still — quite rightly — of great public interest. The second reason for publicity is that in the UK the Health and Safety Executive reports all accidents and near-misses involving radioactivity to the Secretary of State who presents them to Parliament: it is from here, and from announcements by the industry itself, that the Press, radio and TV get their initial information. The extent to which the information is used and the way it is presented and commented upon is something which is decided in the light of editorial policy and journalistic skill.

Coverage in the media may indeed reflect public interest but it can also distort it and make balanced judgements more rather than less difficult to reach. The industry's replies and explanations are almost bound to have less impact and to be less memorable than the news itself. And descriptions of safety precautions, and their undisputed success over the almost whole history of the industry, are not 'news'.

Increase in routine discharges

It is sometimes assumed that routine radioactive waste discharges to the environment will be bound to keep pace with any increase in nuclear power generation or fuel reprocessing capacity, leading to an increase in environmental levels of radioactivity, and eventually to dangerous amounts reaching the human food-chain itself. But the setting of discharge levels is not, as we have seen, left to the discretion of plant operators: it is the job of the regulating authorities, who are independent of the nuclear industry, and they have made it clear, as far as Britain is concerned, that there will be no such relaxation of standards. The operators themselves are constantly improving the efficiency of their clean-up techniques. It is not good enough for discharges to be within the imposed limit: they must be kept as low as is reasonably practicable below those limits and the operators must satisfy the authorities that this is indeed being done.

The long-term waste problem

One of the most frequently heard criticisms of the nuclear industry is that it has still not solved the problem of what to do with its dangerous, highly radioactive wastes. 'The lion of nuclear power has been tamed: the problem now is to keep his cage clean.' Until the industry has shown that it can do this to the satisfaction of its critics, we should (it is said) cut back on the expansion of nuclear power and especially on any commitment to the large-scale use of fast reactors. Otherwise, it is argued, we are laying up trouble for succeeding generations: we, of this generation, will have made use of nuclear energy while consciously leaving the problem of wastes for our children and grandchildren to cope with as best they can.

What the industry has in fact done and is doing has already been described. Work is going ahead in many countries, both singly and jointly, to find ways to 'package' the wastes (e.g. as glass blocks clad in steel) and to dispose of the packages deep in the ground or under the sea where their contents will never again trouble man. Scientists and engineers working on the project are confident that they are well on the way to success in solving the 'packaging' problem and that, within the timescale of perhaps a few decades needed for the glass blocks to cool off sufficiently to allow of burial, the disposal problem will have been solved as well. Technically, it is not particularly difficult, but ironically, one of the major difficulties facing the UK industry at the time of writing is to gain permission for the exploratory drillings that are needed to provide detailed information about rock structures. Local populations, perhaps not surprisingly, tend to regard any proposal for test drilling as the probable forerunner of a full-scale disposal site in the immediate neighbourhood.

Malice aforethought

It is sometimes argued that it is only a question of time before an act of nuclear terrorism or blackmail is perpetrated and, recognising this, we should pull out of the nuclear business before it is too late. Consider, however, the problems facing organisations contemplating the threat of using a 'home-made' nuclear weapon, or a release of radioactivity, in order to gain their ends, whether these be political or material. First they would have to obtain several kilograms of plutonium from some stage in the nuclear fuel cycle, purify it sufficiently, make it into a weapon and place it at a strategic point. All of these stages would be both difficult and dangerous to those undertaking them, even without their having to evade the stringent (and undisclosed) security measures deliberately established to prevent such activities. The threat to release large amounts of radioactive fission-products would present the added difficulty of penetrating the very heavy shielding essential for the safe operation of the plant or transport system. The real potency of the nuclear blackmail threat lies not in the damage that it could actually achieve but in the strong undertones of fear that the very words 'radioactivity' and 'plutonium' have come to carry. Radioactivity has very little to recommend it as an instrument of death and destruction, and the abandonment of nuclear power would not in itself greatly reduce the potential threat of such technological blackmail.

It is sometimes said that the vigilance needed to protect the materials and installations of nuclear power will call for the unacceptable paraphernalia and methods of a police state, but it is of course the prevalence of terrorism rather than the existence of potential terrorist weapons that must govern the direction and size of future developments in security and surveillance.

Proliferation of nuclear weapons

It is also argued that any increase in civil nuclear power is bound to increase the chances of a nuclear war starting somewhere, by placing fissile material in the hands of unscrupulous governments. But this ignores the fact that most industrialised countries could even now develop nuclear weapons if they chose to do so, without going to the trouble and expense of having a nuclear power programme. International accord and political influence still provide much the best means of preventing the proliferation of nuclear weapons, and they can be supplemented if necessary by a variety of physical methods of inspection and monitoring and of 'technical fixes' to make undetected diversion of civil material difficult. Some are already agreed and supervised on a supra-national basis by many countries with existing nuclear power programmes. A world economy that is stagnating through lack of the energy resources needed to keep up a healthy level of trade between nations is a much more likely breeding ground for international discontent, and for the rise of irresponsible, ambitious or frightened governments, than is a prosperous and busy world freed from the threat of energy shortages.

Finally, some people oppose nuclear power for reasons which are frankly political. Such factors form an increasing part of the platform of today's growing 'nuclear debate', but, with their counter-arguments, these have no proper place in the body of this book. It is important, however, to be aware that in the debate technical arguments are often put forward to support or run down views that are themselves basically political: it is important to be able to spot the difference.

Thorium

Some people believe that by adopting a different technical approach we could exploit nuclear energy successfully without running into some of the difficulties that we are faced with at present. In particular, it is suggested that the dangers associated with plutonium could be largely avoided if thorium were used as a fertile material for fast reactors instead of uranium 238.

Thorium has been used experimentally in HTRs and CANDU reactors, and a thorium/uranium 233 cycle is a possibility which might open up reserves of fuel thought to be of the same order of size as those of uranium. (See Chapter 5.) A reactor working on the thorium/U-233 fuel cycle would need an initial charge of enriched uranium before it could become self-sustaining on thorium alone. Fuel fabrication, reprocessing and waste management would probably be no less complex than in the uranium/plutonium cycle, but none of the technology has yet been developed to the point where it can be relied upon for large-scale installation. Some plutonium would still be produced (though not as much as in the present uranium/plutonium cycle) and uranium 233 would present broadly similar problems of safety, security and proliferation to those of plutonium itself. It would be unwise to rely on thorium for any early contribution to world energy needs, and its use should certainly not be thought of as a way of avoiding some of the difficulties associated with the present nuclear fuel cycle.

Nuclear fusion

Some people, mistrustful of nuclear fission and the radioactive wastes, etc., that it inevitably gives rise to, would have us concentrate our efforts on fusion — 'taming the H-bomb reaction' — and, by sheer weight of research, bring it into general use before the energy crisis is really upon us. Let us look at this proposal realistically.

Nuclear fusion, the converse of nuclear fission, is the joining together of light nuclei (particularly the isotopes of hydrogen) to form heavier nuclei (helium) and neutrons, with the liberation of a very large amount of energy. As in fission, this energy comes from the loss in mass arising in the change from small to larger nuclei. This mass loss is greater (in relation to the total masses involved) in fusion than it is in fission, so less fuel is required to produce a given amount of energy. Of several possible nuclear fusion reactions the one between deuterium (hydrogen of mass 2) and tritium (hydrogen of mass 3) shows the greatest promise of early success (see fig. 40). Deuterium is present in sea-water to the extent of about 1 gram in 30 litres. Tritium, which scarcely occurs naturally, would be prepared at the reactor by absorbing the neutrons that arise in the reaction in a surrounding blanket of molten lithium,

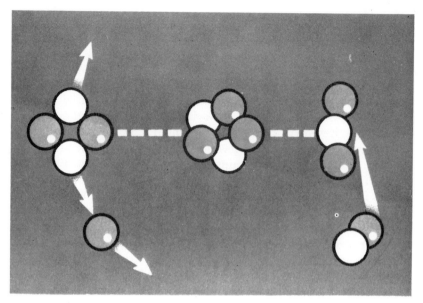

Fig. 40   *Nuclear fusion of deuterium and tritium to form helium and neutron.*

which would also act as coolant taking the heat to the boilers. Lithium is fairly plentiful and it appears unlikely that world supplies would ever run low, but if they did the slightly less effective reaction between deuterium nuclei alone could be used. Only about a tonne of fuel would be needed to supply a 2,000 MW(e) fusion power station for a year, so fuel resources would seem to be assured for Man's foreseeable future.

The technology of achieving fusion is still not proved, but it is generally agreed that for a deuterium–tritium reaction to take place that would yield more energy than is required to initiate it, the reacting materials would have to be at temperatures around 100 million degrees centigrade and they would have to be held together at a concentration of around 10^{14} atoms per cubic centimetre (a modest industrial vacuum) for about one second. Matter at these temperatures consists solely of 'plasma' – nuclei and free electrons moving independently of one another. Being a conductor of electricity, plasma can be influenced by magnetic fields and these can be used to confine it even at these enormous temperatures – at which of course no solids can exist at all as such.

Most attempts at harnessing fusion so far have been directed to achieving the conditions for the reaction by heating the plasma with electric currents while confining it with magnetic fields. Various experimental configurations are used, of which 'toroidal' or motor-tyre shapes are the most favoured (see fig. 41). More recently other lines of approach have received increasing attention, in particular the use of beams of particles from accelerators, and intense pulses of finely focussed laser light. Substantial progress continues to be made towards achieving the conditions set out above.

One of the world's largest experimental fusion devices is the Joint European Torus 'JET', which is now being built at Culham in Oxfordshire by the countries of the European Economic Community, Sweden and Switzerland (see fig. 42). JET is expected to be in operation in the early 1980s and in an experimental programme lasting at least five years will study plasmas in conditions and dimensions approaching those needed for a reactor. In the later stages of the programme actual deuterium–tritium fusion reactions will be studied.

Detailed thought is also being given to the formidable engineering

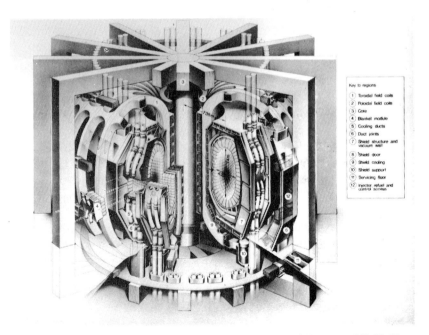

Fig. 41   *Conceptual diagram of the Culham Tokamak Reactor, Mk II. Note the scale: the radius of the toroidal vacuum chamber is 7.38 metres and the longer axis of its bore is 7.4 metres.*

Fig. 42   *A scale model of the vacuum chamber of the Joint European Torus at Culham (JET).*

problems which will have to be solved before a practical fusion reactor can operate. From this it is already clear that a fusion power station will be a very expensive thing: even if the fuel is plentiful the power that it supplies is unlikely to be cheaper than conventional fission power.

If fusion works as well as is hoped it could solve the world's energy resources problems for ever, but it is unlikely in any event to be with us before the early decades of the next century. At present it provides no excuse for letting up on the continued development of nuclear fission power, and particularly of the fast reactor.

Most of the scientists working in the fusion field are optimistic, and the governments of most of the major industrialised countries — including Britain — have sizeable nuclear fusion research programmes, in which there is considerable international co-operation. We know that the fusion reaction will 'go' — the Sun shines, the H-bomb explodes, the particle accelerators produce neutrons: it is a case of finding out how to make it 'go' quietly and to order. If Man can get to the Moon at will, it is hardly likely that power from controlled nuclear fusion will prove to be beyond his eventual capabilities.

Where do we go from here?

We have learned something of the technicalities of nuclear power and its relation to other energy sources, and discussed some of the objections to it. What of the future? Plans need to be made, based on the best available forecasts. Yet forecasts very often go astray and the forecaster gets blamed for the results. Many expert forecasts have been made about the world's energy, some of them by responsible authorities who have a lot to lose if they are wrong, others by people with strong views but little claim to special knowledge. Some of these forecasts may be partly right, all of them will most likely be partly wrong. What follows is the present writer's personal opinion as formed during some twenty-four years on the sidelines of the nuclear industry.

The world, and Britain in particular stands at a point of decision as regards nuclear power: do we continue to develop it to its full potential in spite of its associated hazards and the anxieties they cause (for nobody claims that is has no hazards)? Or do we hold back from increasing our use of it (perhaps even closing down power stations that we are already using) until all its dangers have been overcome? The supporters of this second course consider that the seriousness of the energy supply situation is exaggerated: there is, they say, plenty more oil to be discovered; we have lots of coal; the clean and renewable sources of energy — and of course fusion — are there to be tapped if only we would direct enough of our efforts towards them. In any case, we use far more energy than we should: we ought to cut down on it drastically, even if this leads to interference with traditional freedoms of choice, and a check to growth in the economy. By adopting this approach, it is argued, we would be rid of the problems of radioactivity and its release, the related temptation to terrorists would be removed (and with it the occasion for increased interference with civil liberties) and the dangers of weapons proliferation would be reduced. Our children and grandchildren would not be faced with unsolved, and perhaps insoluble, problems of safeguarding and disposing of nuclear wastes and obsolete nuclear reactors and installations.

In the last analysis this view of the future is based on the belief that nuclear radiations are too dangerous to justify the use of nuclear power. It does not matter whether they arise routinely or accidentally from nuclear reactors and their wastes, or deliberately from the spread of nuclear weapons or the actions of terrorists.

Plutonium, the fuel of fast reactors, is regarded as a wicked substance to be eschewed at all costs.

This approach is negative throughout. The pro-nuclear course is bolder and far more positive. It is based on the belief that a large expansion of nuclear generation is important to man's well-being (as opposed to his mere survival) and essential to the improvement of his lot in the world of today and tomorrow. The attendant drawbacks can be kept well within the limits that we have got used to accepting in other fields. Nuclear power has proved on past experience to be far less demanding on life and health, and in its impact on the environment, than most other forms of energy production. It has also proved cheaper. Other energy sources, including the 'renewables', the supporters of the nuclear case agree, must also be developed to an extent that will depend on the technical and commercial success that their own merits bring. Even taken with energy conservation, however, they cannot be guaranteed to satisfy more than a small fraction of our pressing needs. Though coal is plentiful, coal-fired generation can only expand slowly at best, while oil and gas are quickly becoming too valuable for burning. Nuclear fusion must for the foreseeable future be regarded as a possible first prize to be won in a competition and spending on it should reflect the prospects of winning. Only an orderly and continued expansion in 'conventional' nuclear power can make us sure of meeting the world's needs over the next fifteen to twenty years, while in the longer term only fast reactors can do this. Moreover, far from leading us into an economy based on international dealings in large amounts of plutonium, fast reactors alone are able both to use plutonium as a highly economic form of fuel and to incinerate any surplus. The only wastes that will be left to our children and grandchildren, even from a greatly expanded nuclear power industry, will be a few thousand tons of glassified nuclear wastes safely buried, or awaiting burial, deep beneath the ground or under the ocean bed. Oil and gas will still be available for premium uses and for the Third World, and we will have had time to develop the renewable resources unhurriedly and with the same attention to safety and environmental impact as has characterised the nuclear industry from its very start.

However, nuclear power can do far more than help the world out of its present difficulties. By exploiting uranium, and later perhaps thorium, to the full it can provide a breathing-space in which we may pave the way towards a future for our children and grandchildren in which energy need never again look like running short.

The need for decisions is urgent, particularly in Britain, a small country with a big population relying on the export of its manufactured goods to pay for its food imports and its raw materials. Our engineering capacity is not large, especially in the nuclear field, and without orders it will atrophy and its best young professionals will go abroad. We will be left without the capacity to build nuclear power stations, particularly the fast reactors that provide the key to next century's energy supply, and we will not have the cash to go cap in hand to buy from more advanced countries. We will either have to do without or undertake hurried development that may skimp on safety and will inevitably cost more. It is the writer's firm belief that boldness, ruled by common sense, should and will prevail over timidity, and that the potential of nuclear power will be realised to the full. But if we are going to unlock the atom and release its power to help us before it is too late, we must press ahead now.

FURTHER READING

BECKMAN, P. *The Health Hazards of not going Nuclear* Golden Press, Colorado, 1979

CLARK, R.W. *The Birth of the Bomb* Horizon Press, 1961

FRISCH, O.R. *What Little I Remember* Cambridge University Press, 1979

GOWING, M. *Britain and Atomic Energy 1939–1945* Macmillan, 1964

GOWING, M. *Independence and Deterrence: Britain and Atomic Energy 1945–1952* Macmillan, 1974

GROVES, L.R. *Now It Can be Told: the Story of the Manhattan Project* Harper & Brothers, 1962

HOUSEHOLD, G. *Hostage London* (fiction) Sphere Books, 1977

HOYLE, Sir F. *Energy or Extinction* Heinemann, 1977

JAY, K. E. B. *Nuclear Power: Today and Tomorrow* Spottiswoode Ballantyne & Co, 1961

PATTERSON, W. *Nuclear Power* Penguin Books, 1976

SHERFIELD, Lord (ed.). *Economic and Social Consequences of Atomic Energy* Oxford University Press, 1972

STAPLEDEN, O. *Last and First Men* (fiction) Methuen, 1930

WILSON, J. (ed.) *All in Our Time – The Reminiscences of Twelve Nuclear Pioneers* Bulletin of Atomic Scientists, 1975

UKAEA PUBLICATIONS
Annual reports
Atom (monthly journal)
Glossary of Atomic Terms

INDEX